PROGRESS IN NANOTECHNOLOGY

ENCYCLOPAEDIA OF NANOSCIENCE

Vol. II

PROGRESS IN NANOTECHNOLOGY

By

Dr. S.K. Prasad

School of Studies of Zoology & Biotechnology
Vikram University
Ujjain

DISCOVERY PUBLISHING HOUSE
NEW DELHI-110002

First Published-2007

ISBN 978-81-8356-247-8 (Set)

Published by

DISCOVERY PUBLISHING HOUSE
4831/24, Ansari Road, Prahlad Street,
Darya Ganj, New Delhi-110002 (India)
Phone: 23279245 • Fax: 91-11-23253475
E-mail: dphbooks@rediffmail.com
dphtemp@indiatimes.com

Printed at:

Sachin Printers, Delhi

PREFACE

There has been rapid progress in nanoscience over the last few years, particularly in the area of miniaturization. The present title **Encyclopaedia of Nanoscience** is the first scientifically detailed description of developments that will revolutionize most of the industrial processes and products currently in use. This ground breaking work draws on physics and chemistry to establish basic concepts and analytical tools and thus only it provides an indispensable introduction to the emerging field of Nanoscience, Scientific research is the ultimate tool in pushing forward the limit of understanding. But, as with any tool, research is only powerful if used properly, and to its full effects. The more we learn, the more we discover connections threading through biochemical and biophysical world. In writing this Encyclopaedia the author has made every effort to present these connections in a way that will help first time students of Nanoscience understand the subject and how very relevant it is to their lives.

In the preparation of this book large number of books and research papers have been consulted. So no authenticity is claimed.

The author expresses his gratitude to Mr. Wasan and staff of M/s Discovery Publishing House for their whole hearted co-operation in the publication of this book.

The author tried hard to be accurate and upto date in statement and realises the impossibility of completely avoiding errors therefore, the author will greatly appreciate having his attention called to any questionable statement.

Author

CONTENTS

1

INTRODUCTION

The critical attitude may be described as the conscious attempt to make our theories, our conjectures, suffer in our stead in the struggle for the survival of the fittest. It gives us a chance to survive the elimination of an inadequate hypothesis—when a more dogmatic attitude would eliminate it by eliminating us. As we look forward to see where the technology race leads, we should ask three questions. What is possible, what is achievable, and what is desirable? First, where hardware is concerned, natural law sets limits to the possible. Because assemblers will open a path to those limits, understanding assemblers is a key to understanding what is possible. Second, the principles of change and the facts of our present situation set limits to the achievable. Because evolving replicators will play a basic role, the principles of evolution are a key to understanding what will be achievable. As for what is desirable or undesirable, our differing dreams spur a quest for a future with room for diversity, while our shared fears spur a quest for a future of safety.

These three questions—of the possible, the achievable, and the desirable—frame an approach to foresight. First, scientific and engineering knowledge form a map of the limits of the possible. Though still blurred and incomplete, this map outlines the permanent limits within which the future must move. Second, evolutionary principles determine what paths lie open, and set limits to achievement—including lower limits, because advances that promise to improve life or to further military power will be virtually unstoppable. This allows a limited prediction: If the eons-old evolutionary race does not somehow screech to a halt, then competitive pressures will mold our technological future to the contours of the limits of the possible. Finally, within the broad confines of the possible and the achievable, we can try to reach a future we find desirable.

PREDICTING FUTURE

But how can anyone predict the future? Political and economic trends are notoriously fickle, and sheer chance rolls dice across continents. Even the comparatively steady advance of technology often eludes prediction. Prognosticators often guess at the times and costs required to harness new technologies. When they reach beyond outlining possibilities and attempt accurate predictions, they generally fail. For example, though the space shuttle was clearly possible, predictions of its cost and initial launch date were wrong by several years and billions of dollars. Engineers cannot accurately predict when a technology will be developed, because development always involves uncertainties.

But we have to try to predict and guide development. Will we develop monster technologies before cage technologies, or after? Some monsters, once loosed, cannot be caged. To survive, we must keep control by speeding some developments and slowing others.

Though one technology can sometimes block the dangers of another (defense vs. offense, pollution controls vs. pollution), competing technologies often go in the same direction. On December 29, 1959, Richard Feynman (now a Nobel laureate) gave a talk at an annual meeting of the American Physical Society entitled "There's Plenty of Room at the Bottom." He described a *non-biochemical* approach to *nanomachinery* (working down, step by step, using larger machines to build smaller machines), and stated that the principles of physics do not speak against the possibility of maneuvering things atom by atom. It is not an attempt to violate any laws; it is something, in principle, that can be done; but, in practice, it has not been done because we are too big.... Ultimately, we can do chemical synthesis.... put the atoms down where the chemist says, and so you make the substance." In brief, he sketched another, nonbiochemical path to the assembler. He also stated, even then, that it is "a development which I think cannot be avoided." Assemblers and intelligent machines will simplify many questions regarding the time and cost of technological developments. But questions of time and cost will still muddy our view of the period between the present and these breakthroughs. Richard Feynman saw in 1959 that nanomachines could direct chemical synthesis, presumably including the synthesis of DNA. Yet he could foresee neither the time nor the cost of doing so.

In fact, of course, biochemists developed techniques for making DNA without programmable nanomachines, using shortcuts based on specific chemical tricks. Winning technologies often succeed because of unobvious tricks and details. In the mid-1950s physicists could see that basic semiconductor principles made microcircuits physically possible, but foreseeing how they would be made—foreseeing the details of mask-making, resists, oxide growth, ion implantation, etching, and so forth, in all their complexity—would have been impossible. The nuances of detail and competitive advantage that select winning technologies make the technology race complex and its path unpredictable.

But does this make long-term forecasting futile? In a race toward the limits set by natural law, the finish line is predictable even if the path and the pace of the runners are not. Not human whims but the unchanging laws of nature draw the line between what is physically possible and what is not—no political act, no social movement can change the law of gravity one whit. So however futuristic they may seem, sound projections of technological possibilities are quite distinct from predictions. They rest on timeless laws of nature, not on the vagaries of events.

It is unfortunate that this insight remains rare. Without it, we stumble in a daze across the landscape of the possible, confusing mountains with mirages and discounting both. We look ahead with minds and cultures rooted in the ideas of more sluggish times, when both science and technological competition lacked their present strength and speed. We have only recently begun to evolve a tradition of technological foresight.

WEBBING

Science and technology intertwine. Engineers use knowledge produced by scientists; scientists use tools produced by engineers. Scientists and engineers both work with mathematical descriptions of natural laws and test ideas with experiments. But science and technology differ radically in their basis, methods, and aims. Understanding these differences is crucial to sound foresight. Though both fields consist of evolving meme systems, they evolve under different pressures. Consider the roots of scientific knowledge.

Through most of history, people had little understanding of evolution. This left philosophers thinking that sensory evidence, through reason, must somehow imprint on the mind all human knowledge-including knowledge of natural law. But in 1737, the Scottish philosopher

David Hume presented them with a nasty puzzle: he showed that observations cannot logically prove a general rule, that the Sun shining day after day proves nothing, logically, about its shining tomorrow. And indeed, someday the Sun will fail, disproving any such logic. Hume's problem appeared to destroy the idea of rational knowledge, greatly upsetting rational philosophers (including himself) They thrashed and sweated, and irrationalism gained ground. In 1945, philosopher Bertrand Russell observed that "the growth of unreason throughout the nineteenth century and what has passed of the twentieth is a natural sequel to Hume's destruction of empiricism." Hume's problem-meme had undercut the very idea of rational knowledge, at least as people had imagined it. In recent decades, Karl Popper (perhaps the scientists' favourite philosopher of science), Thomas Kuhn, and others have recognized science as an evolutionary process. They see it not as a mechanical process by which observations somehow generate conclusions, but as a battle where ideas compete for acceptance.

All ideas, as memes, compete for acceptance, but the meme system of science is special: it has a tradition of deliberate idea mutation, and a unique immune system for controlling the mutants. The results of evolution vary with the selective pressures applied, whether among test tube RNA molecules, insects, ideas, or machines. Hardware evolved for refrigeration differs from hardware evolved for transportation, hence refrigerators make very poor cars. In general, replicators evolved for A differ from those evolved for B. Memes are no exception. Broadly speaking, ideas can evolve to seem true or they can evolve to be true (by seeming true to people who check ideas carefully). *Anthropologists* and historians have described what happens when ideas evolve to seem true among people lacking the methods of science; the results (the evil-spirit theory of disease, the lightson—a-dome theory of stars, and so forth) were fairly consistent worldwide. *Psychologists* probing people's naive misconceptions about how objects fall have found beliefs like those that evolved into formal "*scientific*" systems during the Middle Ages, before the work of Galileo and Newton.

Galileo and Newton used experiments and observations to test ideas about *objects* and *motion*, beginning an era of dramatic scientific progress: Newton evolved a theory that survived every test then available. Their method of deliberate testing killed off ideas that strayed too far from the truth, including ideas that had evolved to appeal to the naive human mind. This trend has continued. Further variation and testing have forced the further evolution of scientific ideas, yielding some as bizarre-seeming as the varying time and curved space of relativity, or the probabilistic particle wave functions of quantum mechanics. Even biology has discarded the special life-force expected by early biologists, revealing instead elaborate systems of invisibly small molecular machines. Ideas evolved to be true (or close to the truth) have again and again turned out to seem false—or incomprehensible. The true and the true-seeming have turned out to be as different as cars and refrigerators.

Ideas in the physical sciences have evolved under several basic selection rules. First, scientists ignore ideas that lack testable consequences; they thus keep their heads from being clogged by useless parasites. Second, scientists seek replacements for ideas that have failed tests. Finally, scientists seek ideas that make the widest possible range of exact predictions, The *law of gravity*, for example, describes how stones fall, planets orbit, and galaxies swirl, and makes exact predictions that leave it wide open to disproof. Its breadth and precision likewise give it broad usefulness, helping engineers both to design bridges and to plan spaceflights.

The scientific community provides an environment where such memes spread, forced by competition and testing to evolve toward power and accuracy. Agreement on the importance of testing theories holds the scientific community together through fierce controversies over the theories themselves.

Inexact, limited evidence can never prove an exact, general theory, but it can disprove some theories and so help scientists choose among them. Like other evolutionary processes, science creates something positive (a growing store of useful theories) through a double negative (disproof of incorrect theories). The central role of negative evidence accounts for some of the mental upset caused by science: as an engine of disproof, it can uproot cherished beliefs, leaving psychological voids that it need not refill. In practical terms, of course, much scientific knowledge is as solid as a rock dropped on your toe. We know Earth circles the Sun (though our senses suggest otherwise) because the theory fits endless observations, and because we know why our senses are fooled. We have more than a mere theory that atoms exist: we have bonded them to form molecules, tickled light from them, seen them under *microscopes* (barely), and smashed them to pieces. We have more than a mere *theory of evolution*: we have observed *mutations*, observed *selection*, and observed *evolution* in the laboratory. We have found the traces of past evolution in our planet's rocks, and have observed evolution shaping our tools, our minds, and the ideas in our minds—including the idea of evolution itself. The process of science has hammered out a unified explanation of many facts, including how people and science themselves came to be.

When science finishes disproving theories, the survivors often huddle so close together that the gap between them makes no practical difference. After all, a practical difference between two surviving theories could be tested and used to disprove one of them. The differences among modern theories of gravity, for instance, are far too subtle to trouble engineers who are planning flights through the gravity fields of space. In fact, engineers plan spaceflights using *Newton's disproved theory* because it is simpler than Einstein's, and is accurate enough. *Einstein's theory* of gravity has survived all tests so far, yet there is no absolute proof for it and there never will be. His theory makes exact predictions about everything everywhere (at least about gravitational matters), but scientists can only make approximate measurements of some things somewhere. And, as Karl Popper points out, one can always invent a theory so similar to another that existing evidence cannot tell them apart.

Though media debates highlight the shaky, disputed *borders of knowledge*, the power of science to build agreement remains clear. Where else has agreement on so much grown so steadily and so internationally? Surely not in politics, religion, or art. Indeed, the chief rival of science is a relative: engineering, which also evolves through proposals and rigorous testing.

Evolution of Technology

As IBM Director of Research Ralph E. Gomory says, "The evolution of technology development is often confused with science in the public mind." This confusion muddles our efforts at foresight.

Though engineers often tread uncertain ground, they are not doomed to do so, as scientists are. They can escape the inherent risks of proposing precise, universal scientific theories. Engineers need only show that under particular conditions particular objects will perform well enough. A designer need know neither the exact stress in a suspension bridge cable nor the exact stress that will break it; the cable will support the bridge so long as the first remains below the second, whatever they may be.

Though *measurements* cannot prove precise equality, they can prove inequality. Engineering results can thus be solid in a way that precise scientific theories cannot. Engineering results can even survive disproof of the scientific theories supporting them, when the new theory gives similar results. The case for assemblers, for example, will survive any possible refinements in our theory of quantum mechanics and molecular bonds. Predicting the content of new scientific knowledge is logically impossible because it makes no sense to claim to know already the facts you will learn in the future. Predicting the details of future technology, on the other hand, is merely difficult. Science aims at knowing, but engineering aims at doing; this lets engineers speak of future achievements without paradox. They can evolve their hardware in the world of mind and computation, before cutting metal or even filling in all the details of a design.

Scientists commonly recognize this difference between scientific foresight and technological foresight: they readily make technological predictions about science. Scientists could and did predict the quality of Voyager's pictures of Saturn's rings, for example, though not their surprising content. Indeed, they predicted the pictures' quality while the cameras were as yet mere ideas and drawings. Their calculations used well-tested principles of optics, involving no new science.

Because science aims to understand how everything works, scientific training can be a great aid in understanding specific pieces of hardware. Still, it does not automatically bring engineering expertise; designing an airliner requires much more than a knowledge of the sciences of metallurgy and aerodynamics. Scientists are encouraged by their colleagues and their training to focus on ideas that can be tested with available apparatus. The resulting short-term focus often serves science well: it keeps scientists from wandering off into foggy worlds of untested fantasy, and swift testing makes for an efficient mental immune system. Regrettably, though, this cultural bias toward short-term testing may make scientists less interested in long-term advances in technology.

The impossibility of genuine foresight regarding science leads many scientists to regard all statements about future developments as "*speculative*"—a term that makes perfect sense when applied to the future of science, but little sense when applied to well-grounded projections in technology. But most engineers share similar leanings toward the short term. They too are encouraged by their training, colleagues, and employers to focus on just one kind of problem: the design of systems that can be made with present technology or with technology just around the corner. Even long-term engineering projects like the space shuttle must have a technology cutoff date after which no new developments can become part of the basic design of the system.

In brief, scientists refuse to predict future scientific knowledge, and seldom discuss future engineering developments. Engineers do project future developments, but seldom discuss any not based on present abilities. Yet this leaves a crucial gap: what of engineering developments firmly based on present science but awaiting future abilities? This gap leaves a fruitful area for study.

Imagine a line of development which involves using existing tools to build new tools, then using those tools to build novel hardware (perhaps including yet another generation of tools). Each set of tools may rest on established principles, yet the whole development sequence may take many years, as each step brings a host of specific problems to iron out. Scientists planning their next experiment and engineers designing their next device may well ignore all but the first step. Still, the end result may be foreseeable, lying well within the bounds of the possible shown by established science. Recent history illustrates this pattern. Few engineers considered building space stations before rockets reached orbit, but the principles were clear enough, and space systems engineering is now a thriving field. Similarly, few mathematicians and engineers studied the possibilities of computation until computers were built, though many did afterward. So it is not too surprising that few scientists and engineers have yet examined the future of nanotechnology, however important it may become.

Lesson of Leonardo

Efforts to project engineering developments have a long history, and past examples illustrate present possibilities. For example, how did Leonardo da Vinci succeed in foreseeing so much, and why did he sometimes fail?

Leonardo lived five hundred years ago, his life spanning the discovery of the New World. He made projections in the form of drawings and inventions; each design may be seen as a projection that something much like it could be made to work. He succeeded as a mechanical engineer: he designed workable devices (some were not to be built for centuries) for excavating, metalworking, transmitting power, and other purposes. He failed as an aircraft engineer: we now know that his flying machines could never be made to work as described.

His successes at machine design are easy to understand. If parts can be made accurately enough, of a hard enough, strong enough material, then the design of slow-moving machines with levers, pulleys, and rolling bearings becomes a matter of geometry and leverage. Leonardo understood these quite well. Some of his "*predictions*" were long-range, but only because many years passed before people learned to make parts precise enough, hard enough, and strong enough to build (for instance) good ball bearings—their use came some three hundred years after Leonardo proposed them. Similarly, gears with superior, cycloidal teeth went unmade for almost two centuries after Leonardo drew them, and one of his chaindrive designs went unbuilt for almost three centuries.

His failures with aircraft are also easy to understand. Because Leonardo's age lacked a science of *aerodynamics*, he could neither calculate the forces on wings nor know the requirements for aircraft power and control. Can people in our time hope to make projections regarding molecular machines as accurate as those Leonardo da Vinci made regarding metal machines? Can we avoid errors like those in his plans for *flying machines*? Leonardo's example suggests that we can. It may help to remember that Leonardo himself probably lacked confidence in his aircraft, and that his errors nonetheless held a germ of truth. He was right to believe that flying machines of some sort were possible-indeed, he could be certain of it because they already existed. Birds, bats, and bees proved the possibility of flight. Further, though there were no working examples of his ball bearings, gears, and chain drives, he could have confidence in their principles. Able minds had already built a broad foundation of knowledge about geometry and the laws of leverage. The required strength and accuracy of the parts may have caused him doubt, but not their interplay of function and motion. Leonardo could propose machines requiring better parts than any then known, and still have a measure of confidence in his designs.

Proposed molecular technologies likewise rest on a broad foundation of knowledge, not only of geometry and leverage, but of *chemical bonding*, *statistical mechanics*, and physics in general. This time, though, the problems of material properties and fabrication accuracy do not arise in any separate way. The properties of atoms and bonds are the material properties, and atoms come prefabricated and perfectly standardized. Thus we now seem better prepared for foresight than were people in Leonardo's time: we know more about molecules and controlled bonding than they knew about steel and precision machining.

In addition, we can point to nanomachines that already exist in the cell as Leonardo could point to the machines (birds) already flying in the sky. Projecting how second-generation nanomachines can be built by protein machines is surely easier than it was to project how precise steel machines would be built starting with the cruder machines of Leonardo's time. Learning to use crude machines to make more precise machines was bound to take time, and the methods were far from obvious. *Molecular machines*, in contrast, will be built from identical prefabricated atomic parts which need only be assembled. Making precise machines with crooked machines must have been harder to imagine then than molecular assembly is now. And besides, we know that molecular assembly happens all the time in nature. Again, we have firmer grounds for confidence than Leonardo did. In Leonardo's time, people had scant knowledge of electricity and magnetism, and knew nothing of molecules and quantum mechanics. Accordingly, electric lights, radios, and computers would have baffled them. Today, however, the basic laws most important to engineering—those describing normal matter—seem well understood.

As with surviving theories of gravity, the scientific engine of disproof has forced surviving theories of matter into close agreement. Such knowledge is recent. Before this century people did not understand why solids were solid or why the Sun shone. Scientists did not understand the laws that governed matter in the ordinary world of molecules, people, planets, and stars. This is why our century has sprouted transistors and hydrogen bombs, and why molecular technology draws near. This knowledge brings new hopes and dangers, but at least it gives us the means to see ahead and to prepare. When the basic laws of a technology are known, future possibilities can be foreseen (though with gaps, or Leonardo would have foreseen mechanical computers). Even when the basic laws are poorly known, as were the principles of aerodynamics in Leonardo's time, nature can demonstrate possibilities. Finally, when both science and nature point to a possibility, these lessons suggest that we take it to heart and plan accordingly.

Assembler Breakthrough

The foundations of science may evolve and shift, yet they will continue to support a steady, growing edifice of engineering knowhow. Eventually, assemblers will allow engineers to make whatever can be designed, sidestepping the traditional problems of materials and fabrication. Already, approximations and computer models allow engineers to evolve designs even in the absence of the tools required to implement them. All this will combine to permit foresight—and something more.

As nanotechnology advances, there will come a time when assemblers become an *imminent prospect*, backed by an earnest and well-funded development program. Their expected capabilities will have become clear. By then, computer-aided design of molecular systems—which has already begun—will have grown common and sophisticated, spurred by advances in computer technology and the growing needs of *molecular engineers*. Using these design tools, engineers will be able to design secondgeneration *nanosystems*, including the second-generation *assemblers* needed to build them. What is more, by allowing enough margin for inaccuracies (and by preparing alternative designs), engineers will be able to design many systems that will work when first built—they will have evolved sound designs in a world of simulated molecules.

Consider the force of this situation: under development will be the greatest production tool in history, a truly general fabrication system able to make anything that can be designed—and a design system will already be in hand. Will everyone wait until assemblers appear before planning how to use them? Or will companies and countries respond to the pressures of opportunity and competition by designing nanosystems in advance, to speed the exploitation of assemblers when they first arrive? This design-ahead process seems sure to occur; the only question is when it will start and how far it will go. Years of quiet design progress may well erupt into hardware with unprecedented suddenness in the wake of the assembler breakthrough. How well we design ahead —and what we design—may determine whether we survive and thrive, or whether we obliterate ourselves.

Because the assembler breakthrough will affect almost the whole of technology, foresight is an enormous task. Of the universe of possible mechanical devices, Leonardo foresaw only a few. Similarly, of the far broader universe of future technologies, modern minds can foresee only a few. A few advances, however, seem of basic importance. *Medical technology*, the space frontier, *advanced computers*, and new *social inventions* all promise to play interlocking roles. But the assembler breakthrough will affect all of them, and more.

2

FUNCTIONAL NANOTECHNOLOGY

The word *manufacturing* comes from the Latin *manufactus*, meaning "handmade." Today, the term brings to mind huge, noisy machines stamping out products and spewing waste. Giving up manufactured products isn't popular or practical—almost everything we use today is manufactured. If all machine-made products were to suddenly vanish, most people in the United States would find themselves naked and outdoors, with very little around them. Expanding manufacturing is an objective of nearly every nation on Earth. *Molecular manufacturing* can help us get what we seem to want: high-quality products made at low cost with little environmental impact.

What follows is an exploration of the possible—a look at the devices that could be built once precise molecular control is achieved, and a look at how people might run a manufacturing business based on *nanomachines*. Try not to think of these sketches as hard-and-fast predictions of precisely how things will be done, but instead as descriptions of capabilities—the sorts of things that can be done once *nanotechnology* is well in hand. Doubtless there will be better ways to do things than the ways we describe. As usual, references to the 1980s and before are historically accurate; in the rest, the science isn't fiction.

DESERT ROSE INDUSTRIES

Desert Rose Industries is a diversified wholesale manufacturer of enough furniture, computers, toys, and recreation equipment to have made any twentieth-century captain of industry proud. But if you assembled all Desert Rose employees in front of corporate headquarters, you'd see Carl and Maria Santos standing beside a building the size of a four-bedroom house. This industrial giant is a typical mom-and-pop business, helped along by a network of telecommuters who handle sales and customer support from homes scattered across North America.

Their friends chide Carl and Maria as "old-fashioned traditionalists" and tease Maria about abandoning Carl in the factory while she travels to Europe, Asia, South America, and Africa for new business. In the molecular-manufacturing business, familiar personal skills and virtues—honesty, accuracy, good communication—are as important as before. Maria likes to work with the customers. Aided by her S.B. in molecular manufacturing from MIT and her MFA in design, she patiently helps nervous new designers through their first manufacturing experience, and with unflagging courtesy and good humor, handles rush orders, last-second changes, and special orders. Maria's good design ideas and caring personality won them a reputation for being responsive to customer needs. Carl, precise and careful, built their name for accurate manufacturing and delivery on schedule.

Except for Carl's habit of playing Gershwin at full volume with the windows open, the only sounds at the Desert Rose site are the birds along the banks of the stream that winds across the canyon floor; no clanking machinery here. Maria's parents built Desert Rose Industries out here on an old smelter many miles away from human neighbours. They regraded the land and cleaned up the wastes. Maria adapted a molecular processor to convert heavy-metal contaminants back into stable minerals, and shipped them off to help refill the hole they had originally come from, an old open-pit mine. The desert has mostly healed now, and a few tough trees are spreading along the stream again.

New customers coming up the road for a firsthand look at the manufacturing operations get the full tour: a lunch/meeting room, Maria's office, the manufacturing plant, and the warehouse space for parts and products out back. "The plant" is the largest room, and Carl's pride. Twelve manufacturing ponds and their cooling systems—vats ranging in size from a kitchen sink to a small swimming pool—are where Desert Rose uses *nanocomputers* and *assemblers* to do their building work. A plumbers' nightmare of piping runs between the ponds and a triple row of containers with labels like *carbon feedstock*, *prepared platinum*, *size-4 structural fibers*, and *prefab motors*. Carl keeps a good stock of parts and raw materials on hand, with more in the underground warehouse. Sure, some rare things almost never get used, but having them ready to go is one of Carl's secrets for delivering on time and building precisely to specification. Over on a table are Carl's music system and the computers—descendants of the IBM PCs and Macintoshes of the 1980s—that are used to run the manufacturing process. In a space the size of a large living room, Carl and Maria have all the raw materials and all the production equipment—nanocomputers and assemblers—they need for building almost anything.

Occasionally, Carl and Maria need the services of specialized tools, such as *disassemblers*, that might exist only in labs. A disassembler works like an archeologist, painstakingly excavating the structure of a *molecule*, removing *atom* after atom, in order to record and analyze the molecular structure. Because they work so slowly, noting the position of each molecule, disassemblers aren't used for recycling operations—it would be expensive and pointless to record all this unwanted data. But as tools for analyzing the unknown, they're hard to beat.

Maria found this out when a customer sent her an order for tropically scented furniture and fixtures for his restaurant, but instead of including the software instructions for building the perfume, Maria found a plastic bag full of resinous brown gook with a note saying, "I got this stuff in the tropics. Please make the fabric smell like this." Maria (after sniffing the gook and deciding it smelled surprisingly tropically good) shipped the sample to the lab for chemical analysis by disassembler. The lab sent back software with the molecular description and instructions for building the same scent into the furniture.

Carl usually schedules production very tightly: in every tank, assemblers are building products; every computer is directing work. But this morning, listening to the tone of Maria's voice wafting in from the front office, Carl changes his plans: something important is about to happen. He postpones building orders for video wallpaper and commemorative diamond baseballs, and holds three pools and a computer ready. Minutes later, Maria hurries in, her voice tight and anxious. "Carl, that earthquake down south—they need help. Amanda from the Red Cross is sending the software right now."

To build a product, Desert Rose needs design instructions—computer software—for the assemblers. Carl and Maria have their own software library, but usually they buy or rent what they need, or the customers send their own designs.

The software that Amanda sends contains the specifications to manufacture the emergency equipment: a set of instructions to be run on a standard desktop computer. Within minutes, two copies of the Red Cross software arrive electronically. Before starting the build, Carl meticulously checks to make sure that the master copy and backup copy agree and weren't damaged in transit. If the instructions are complete and correct and properly signed with the Red Cross data stamp, then the desktop computer will communicate these building instructions directly to millions of small computers acting as on-the-job foremen directing the work: nanocomputers.

NANOCOMPUTERS

While the first, primitive assemblers were controlled by changing what molecules are in the solution around the device, getting the speed and accuracy wanted for large-scale manufacturing takes real computation. Carl's setup uses a combination of special-purpose molecule processors and general-purpose assemblers, all controlled and orchestrated by nanocomputers.

Computers back in the 1990s used microelectronics. They worked by moving electrical charge back and forth through conducting paths—wires, in effect—using it to block and unblock the flow of charge in other paths. With nanotechnology, computers are built from *molecular electronics*. Like the computers of the 1990s, they use electronic signals to weave the patterns of digital logic. Being made of molecular components, though, they are built on a much smaller scale than 1990s computers, and work much faster and more efficiently. On the scale of our simulated molecular world, 1990s computer chips are like landscapes, while nanocomputers are like individual buildings. Carl's desktop PC contains over a trillion nanocomputers, enough to out-compute all the microelectronic computers of the twentieth century put together.

Back in the dark ages of the 1980s, an exploratory engineer proposed that nanocomputers could be mechanical, using sliding rods instead of moving electrons as shown in Figure 8. These molecular mechanical computers were much easier to design than molecular electronic computers would have been. They were a big help in getting some idea of what nanotechnology could do.

Even back then, it was pretty obvious that mechanical computers would be slower than electronic computers. Carl's molecular electronic PC would have been no great surprise, though nobody knew just how to design one. When nanotechnology actually arrived and people started competing to build the best possible computers, molecular electronics

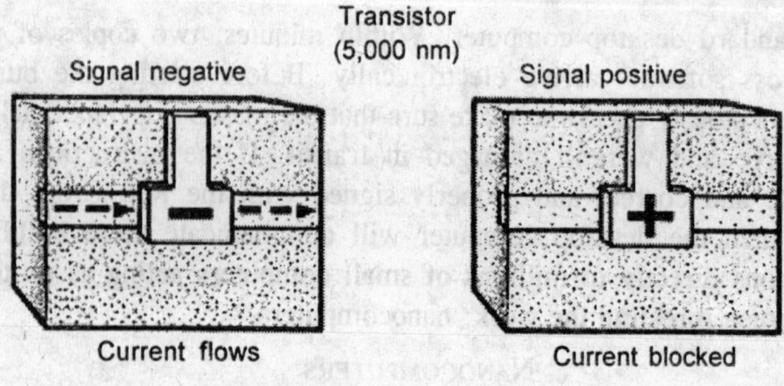

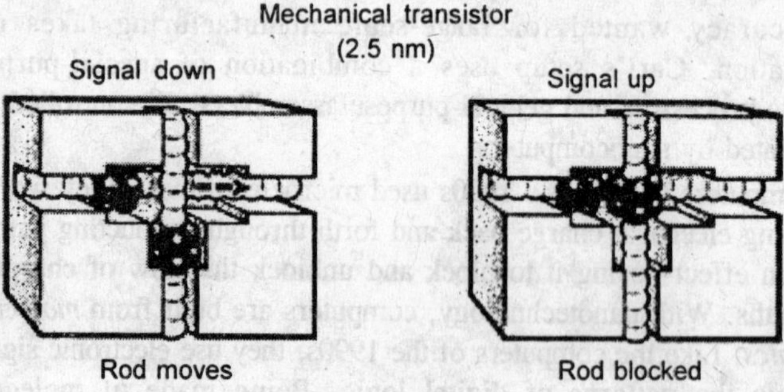

Fig. 2.1. Mechanical transistor.

won the technology race. Still, mechanical nanocomputers could have done all the nanocomputing jobs at Desert Rose: ordinary, everyday molecular manufacturing just doesn't demand the last word in computer performance.

For Carl, the millions of nanocomputers in the milky waters of his building ponds are just extensions of machines on his desk, machines there to help him run his business and deliver products to his customers—or, in the case of the Red Cross emergency, to help provide time-critical emergency supplies. By reserving those three separate ponds, Carl can either build three different kinds of equipment for the Red Cross or use all the ponds to mass-produce the first thing on the Red Cross list: emergency shelters for ten thousand people. The software is ready, the plumbing is fine, the drums of building materials are all topped up, the Special Mix for this job is loaded: the build is ready to start. "Okay," Carl tells the computer, "build Red Cross tents." Computer talks to nanocomputers. In all three pools, nanocomputers talk to assemblers. The build begins.

Assembling Products

Some of the building done at Desert Rose Industries uses assemblers much like the ones we saw in the first hall of the plant tour, back in the simulated molecular world of the Silicon Valley Faire. As seen in simulation, they are big, slow, computer-controlled things moving molecular tools. With the right instructions and machinery to keep them supplied with molecular tools, these general-purpose assemblers can build almost anything. They're slow, though, and take a lot of energy to run. Some of the building uses special-purpose assembly systems in the molecule-processing style, like the systems in the basement we saw in the tour of a simulated molecular factory. The special-purpose systems are all moving belts and rollers, but no arms. This is faster and more efficient, but for quantity orders, cooling requirements limit the speed.

It's faster to use larger, prefabricated building blocks. Desert Rose uses these for most of their work, and especially for rush orders like the one Carl just set up. Their underground warehouse has room-sized bins containing upward of a thousand tons of the most popular building blocks, things like structural fibers. They're made at plants on the West Coast and shipped here by subway for ready use. Other kinds are made on site using the special-purpose assemblers. Carl's main room has several cabinet-sized boxes hooked up to the plumbing, each taking in raw materials, running them through this sort of specialized *molecular machinery*, and pumping out a milky syrup of product. One syrup contains motors, another one contains computers, and another is full of microscopic plug-in light sources. All go into tanks for later use.

Now they're being used. The mix for the Red Cross tent job is mostly structural fiber stronger than the old bulletproof-vest materials. Other building blocks also go in, including motors, computers, and dozens of little struts, angle brackets, and doohickies. The mix would look like someone had stirred together the parts from a dozen toy sets, if the parts were big enough to see. In fact, though, the largest parts would be no more than blurry dots, if you saw one under a normal optical microscope.

The mix also contains block-assemblers, floating free like everything else. These machines are big, about like an office building in our simulation view with the standard settings. Each has several jointed arms, a computer, and several plugs and sockets. These do the actual construction work.

To begin the build, pumps pour the mix into a manufacturing pond. The constant tumbling motions of microscopic things in liquids would be too disorganized for building anything so large as a tent, so the block-assemblers start grabbing their neighbours. Within moments, they have linked up to form a framework spread through the liquid. Now that they are plugged together, they divide up jobs, and get to work. Instructions pour in from Carl's desktop computer.

The block-assemblers use sticky grippers to pull specific kinds of building blocks out of the liquid. They use their arms to plug them together. For a permanent job, they would be using blocks that bond together chemically and permanently. For these temporary tents, though, the Red Cross design uses a set of standard blocks that are put together with amazingly ordinary fasteners: these blocks have snaps, plugs, and screws, though of course the parts are atomically perfect and the threads on the screws are single helical rows of atoms. The resulting joints weaken the tent's structure somewhat, but who cares? The basic materials are almost a hundred times stronger than steel, so there is strength to waste if it makes manufacturing more convenient.

Fiber segments snap together to make fabrics. Some segments contain motors and computers, linked by fibers that contain power and data cables. Struts snap together with more motors and computers to make the tent's main structures. Special surfaces are made of special building blocks. From the human perspective, each tent is a lightweight structure that contains most of the conveniences and comforts of an apartment: cooking facilities, a bathroom, beds, windows, air conditioning, specially modified to meet the environmental demands of the quake-stricken country. From a builder's perspective, especially from a nanomachine's point of view, the tent is just structure slapped together from a few hundred kinds of prefab parts.

In a matter of seconds, each block-assembler has put together a few thousand parts, and its section of the tent is done. In fact, the whole thing is done: many trillions of hands make light work. A crane swings out over the pond and starts plucking out tent packages as fresh mix flows in.

Maria's concern has drawn her back to the plant to see how the build is going. "It's coming along," Carl reassures her. "Look, the first batch of tents is out." In the warehouse, the first pallet is already stacked with five layers of dove-gray "suitcases": tents dried and packed for transport. Carl grabs a tent by the handle and lugs it out the door. He pushes a tab on the corner labeled "Open," and it takes over a minute to unfold to a structure a half-dozen paces on a side. The tent is big, and light enough to blow away if it didn't cling to the ground so tightly. Maria and Carl tour the tent, testing the appliances, checking the construction of furniture: everything is extremely lightweight compared to the bulk-manufactured goods of the 1990s, tough but almost hollow.

Like the other structures, the walls and floors are full of tiny motors and struts controlled by simple computers like the ones used in twentieth-century cars, televisions, and pinball machines. They can unfold and refold. They can also flex to produce sound like a high-quality speaker, or to absorb sound to silence outdoors racket. The whole three-room setup is small and efficient, looking like a cross between a boat cabin and a Japanese business hotel room. Outside, though, it is little more than a box. Maria shakes her head, knowing full well what architects can do these days when they try to make a building really fit its site.

Behind the Scenes and Afterward

Desert Rose Industries and other manufacturers can make almost anything quickly and at low cost. That includes the tunneling machines and other equipment that made the subway system they use for shipping. Digging a tunnel from coast to coast now costs less than digging a single block under New York City used to. It wasn't expensive to get a deep-transit terminal installed in their basement. Just as the tents aren't mere bundles of canvas, these subways aren't slow things full of screeching, jolting metal boxes. They're magnetically levitated to reach aircraft speeds—as experimental Japanese trains were in the late 1980s—making it easy for Carl and Maria to give their customers quick service. There's still a road leading to the plant, but nobody's driven a truck over it for years.

They only take in materials that they will eventually ship out in products, so there's nothing left over, and no wastes to dump. One corner of the plant is full of recycling equipment. There are always some obsolete parts to get rid of, or things that have been damaged and need to be reworked. These get broken down into simpler molecules and put back together again to make new parts.

The gunk in the manufacturing ponds is water mixed with particles much finer than silt. The particles—fasteners, computers, and the rest—stay in suspension because they are wrapped in molecular jackets that keep them there. This uses the same principle as detergent molecules, which coat particles of oily dirt to float them away.

Though it wouldn't be nutritious or appetizing, you could drink the tent mix and be no worse for it. To your body, the parts and their jackets, and even the nanomachines, would be like so many bits of grit and sawdust. (Grandma would have called it roughage.)

Carl and Maria get their power from solar cells in the road, which is the only reason they bothered having it paved. In back of their plant stands what looks like a fat smokestack. All it produces, though, is an updraft of clean, warm air. The darkly paved road, baking in the New Mexico sun, is cooler than you might expect: it soaks up solar energy and makes electricity, instead of just heat. Once the power is used, it turns back into heat, which has to go somewhere. So the heat rises from their cooling tower instead of the road, and the energy does useful work on the way.

Some products, like rocket engines, are made more slowly and in a single piece. This makes them stronger and more permanent. The tents, though, don't need to be superstrong and are just for temporary use. A few days after the tents go up, the earthquake victims start to move out into new housing (permanent, better-looking, and very earthquake resistant). The tents get folded and shipped off for recycling.

Recycling things built this way is simple and efficient: nanomachines just unscrew and unsnap the connectors and sort the parts into bins again. The shipments Desert Rose gets are mostly recycled to begin with. There's no special labeling for recycled materials, because the molecular parts are the same either way.

For convenience (and to keep the plant small), Carl and Maria get most of their parts prefabricated, even though they can make almost anything. They can even make more production equipment. In one of their manufacturing ponds, they can put together a new cabinet full of special-purpose assemblers. They do this when they want to make a new type of part in-house. Like parts, the part-assemblers are made by special-purpose assemblers. Carl can even make big vats in medium-size vats, unfolding them like tents.

If Desert Rose Industries needed to double capacity, Carl and Maria could do it in just a few days. They did this once for a special order of stadium sections. Maria got Carl to recycle the new building before its shadow hurt their cactus garden.

Factory Factories

In the Desert Rose Industries scenario, manufacturing has become cheap, fast, clean, and efficient. Using fast, precise machines to handle matter in molecular pieces makes it easy for nanotechnology to be fast, clean, and efficient. But for it to be *cheap*, the manufacturing equipment has to be cheap.

The Desert Rose scenario shows how this can work. Molecular-manufacturing equipment can be used to make all the parts needed to build more molecular manufacturing equipment. It can even build the machines needed to put the parts together. This resembles an idea developed by NASA for a self-expanding manufacturing complex on the Moon, but made faster and simpler using molecular machines and parts.

Replicators

In the early days of nanotechnology, there won't be as many different kinds of machines as there are at Desert Rose. One way to build a lot of molecular manufacturing equipment in a reasonable time would be to make a machine that can be used to make a copy of itself, starting with special but simple chemicals. A machine able to do this is called a "*replicator.*" With a replicator and a pot full of the right fuel and raw materials, you could start with one machine, then have two, four, eight, and so on.

This doubling process soon makes enough machines to be useful. The replicators—each including a computer to control it and a general-purpose assembler to build things—could then be used to make something else, like the tons of specialized machines needed to set up a Desert Rose manufacturing plant. At that point, the replicators could be discarded in favour of those more efficient machines.

Replicators are worth a closer look, though, because they show how quickly molecular manufacturing systems can be used to build more manufacturing equipment. If we were in one of our standard simulation views, the submicroscopic device at the top of the picture would be like a huge tank, three stories tall when lying on its side. Most of its interior is taken up by a tape memory system that tells how to move the arm to build all the parts of the replicator, except the tape itself. The tape gets made by a special tape-copying machine. At the right-hand end of the replicator are pores for bringing in fuel and raw-material molecules, and machinery for processing them. In the middle are computer-controlled arms, like the ones we saw on the plant trip. These do most of the actual construction.

The steps in the cycle—using a copy to block the tube, beginning a fresh copy, then releasing the old one—illustrate one way for a machine to build a copy of itself while floating in a liquid, yet doing all its construction work inside, in vacuum. (It's easier to design for vacuum, and this is exploratory-engineering work, so easier design is

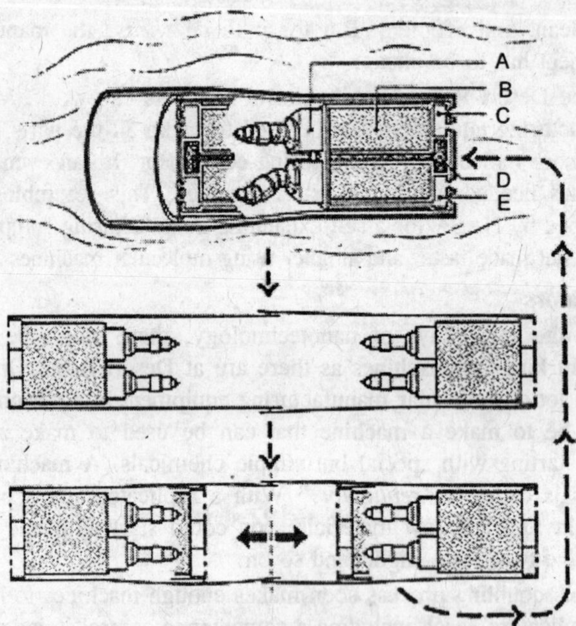

Fig. 2.1. Replicator.

better design.) Calculations suggest that the whole construction cycle can be completed in less than a quarter hour, since the replicator contains about a billion atoms, and each arm can handle about a million atoms per second. At that rate, one device can double and double again to make trillions in about ten hours.

Each replicator just sits in a chemical bath, soaking up what it needs and making more replicators. Eventually, either the special chemicals run out or other chemicals are added to signal them to do something else. At that point, they can be reprogrammed to produce anything else you please, so long as it can be extruded from the front. The products can be long, and can unfold or be pieced together to make larger objects, so the size of these initial replicators–smaller than a bacterium–would be only a temporary limitation.

General Assemblers

From the *molecular manipulators* and primitive assemblers has already been described, the most likely path to nanotechnology leads to assemblers with more and more general capabilities. Still, efficiency favours special-purpose machines, and the Desert Rose scenario didn't make much use of general assemblers. Why bother making general-purpose assemblers in the first place? To see the answer, turn the question around and ask, Why not build such a tool? There is nothing outstandingly difficult about a general assembler, as molecular machinery goes. It will just be a device with good, flexible positional control and a system to feed it a variety of molecular tools. This is a useful, basic capability. General-purpose assemblers could always be replaced by a lot of specialized devices, but to build those specialized devices in the first place, it makes sense to come up with a more flexible, general-purpose system that can just be reprogrammed.

So, general purpose machines are likely to find use in making short production runs of more specialized devices. *Ralph Merkle*, a computers and security expert at Xerox Palo Alto Research Center, sees this as paralleling the way manufacturing works today: "General purpose devices could do many tasks, but they'll do them inefficiently. For any given task, there will be one or a few best ways of doing it, and one or a few special purpose devices that are finely tuned to do that one task. Nails aren't made by a general-purpose machine shop, they're made by nail-making machines. Making nails with a general-purpose machine shop would be more expensive, more difficult, and more time-consuming. Likewise, in the future we won't see a proliferation of general-purpose self-replicating systems, we'll see specialization for almost every task."

What Will These Capabilities Make Possible?

We've surveyed a lot of devices: assemblers of various flavors, nanocomputers, disassemblers, replicators, and others. What's important about these is not the exact distinctions between them, but the capabilities that they will give and the effects they will have on human lives. Again, we are suspending discussion of potential misapplications until later. If we tease apart the implications of what we've seen in the Desert Rose scenario, we can analyze some of the key impacts of molecular manufacturing in industry, science, and medicine.

Technology and Industry

At its base, nanotechnology is about molecular manufacturing, and manufacturing is the basis of much of today's industry. This is why Desert Rose made a good starting point for describing the possibilities of a nanotechnological world. From an industrial perspective, it makes sense to think of nanotechnology in terms of products and production.

New Products

Today, we handle matter crudely, but nanotechnology will bring thorough control of the structure of matter, the ability to build objects to atom-by-atom specifications. This means being able to make almost anything. By comparison, even today's range of products will feel very limited. Nanotechnology will make possible a huge range of new products, a range we can't envision today. Still, to get a feel for what is possible, we can look at some easily imagined applications.

Reliable Products

Today, products often fail, but for failures to occur—for a wing to fall off an airplane, or a bearing to wear out—a lot of atoms have to be out of place. In the future, we can do better. There are two basic reasons for this: better materials and better quality control, both achieved by molecular manufacturing. By using materials tens of times stronger than steel, as Desert Rose did, it will be easy to make things that are very strong, with a huge safety margin. By building things with atom-by-atom control, flaws can be made very rare and extremely small—nonexistent, by present standards. With nanotechnology, we can design in big safety margins and then manufacture the design with near-perfection. The result will be products that are tough and reliable.

Intelligent Products

Today, we make most things from big chunks of metal, wood, plastic, and the like, or from tangles of fibers. Objects made with molecular manufacturing can contain trillions of microscopic motors and computers, forming parts that work together to do something useful. A climber's rope can be made of fibers that slide around and reweave to eliminate frayed spots. Tents can be made of parts that slide and lock to turn a package into a building. Walls and furniture can be made to repair themselves, instead of passively deteriorating. On a mundane level, this sort of flexibility will increase reliability and durability. Beyond this, it will make possible new products with abilities we never imagined we needed so badly. And beyond even this, it will open new possibilities for art.

Inexpensive Production

Today, production requires a lot of labour, either for making things or for building and maintaining machines that make things. Labour is expensive, and expensive machines make automation expensive, too. In the Desert Rose scenario, we got a glimpse of how molecular manufacturing can make production far less expensive than it is today. This is perhaps the most surprising conclusion about nanotechnology, so we'll take a closer look at it in the next chapter.

Clean Production

Today, our manufacturing processes handle matter sloppily, producing pollution. One step puts stuff where it shouldn't be; the next washes it off the product and into the water supply. Our transportation system worsens the problem as unreliable trucks and tankers spill noxious chemicals over the land and sea. Everything is expensive, so companies skimp on even the half-effective pollution controls that we know how to build. Nanotechnology will mean greater control of matter, making it easy to avoid pollution. This means that a little public pressure will go a long way toward a cleaner environment. Likewise, it will make it easy to increase efficiency and reduce resource requirements.Products, like the Red Cross tents at Desert Rose, can be made of snap-together, easily recyclable parts.

Sophisticated products could even be made from biodegradable materials. Nanotechnology will make it easy to attack the causes of pollution at their technological root. Nanotechnology will have great applications in the field of industry, much as transistors had great applications in the field of vacuum tube electronics, and democracy had great applications in the field of monarchy. It will not so much *advance* twentieth-century industry as *replace* it—not all at once, but during a thin slice of historical time.

SCIENCE

Chemistry

Today, chemists work with huge number of molecules and study them using clever, indirect techniques. Making a new molecule can be a major project, and studying it can be another. Molecular manufacturing will help chemists make what they want to study, and it will help them make the tools they need to study it. Nanoinstruments will be used to prod, measure, and modify molecules in a host of ways, studying their structures, behaviours, and interactions.

Materials

Today, materials scientists make new superconductors, semi-conductors, and structural materials by mixing and crushing and baking and freezing, and so forth. They dream of far more structures than they can make, and they stumble across more things than they plan. With molecular manufacturing, materials science can be much more systematic and thorough. New ideas can be tested because new materials can be built according to plan (rather than playing around, groping for a recipe). This need not rule out unexpected discoveries, since experiments—even blind searches—will go much faster. A few tons of raw materials would be enough to make a billion samples, each a cubic micron in size. In all of history so far, materials scientists have never tested so many materials. With nanoinstruments and nanocomputers, they could. One laboratory could then do more than all of today's materials scientists put together.

Biology

Today, biologists use a host of molecular devices borrowed from biology to study biology. Many of these can be viewed as molecular machines. Nanotechnology will greatly advance biology by providing better molecular devices, better nanoinstruments. Some *cells* have already been mapped in amazing molecular detail, but biology still has far to go. With nanoinstruments (including molecule-by-molecule disassemblers), biologists will at last be able to map cells completely and study their interactions in detail. It will become easy not only to find molecules in cells, but to learn what they do. This will help in understanding disease and the molecular requirements for health, enormously advancing medicine.

Computation

Today, computers range from a million to a billion times faster than an old desktop adding machine, and the results have been revolutionary for science. Every year, more questions can be answered by calculations based on known principles of physics. The advent of nanocomputers—even slow, miserable, mechanical nanocomputers—will give us practical machines with a *trillion* times the power of today's computers (essentially by letting us package a trillion computers in a small space, without gobbling too much money or energy.) The consequences will again be revolutionary.

Physics

The known principles of physics are adequate for understanding molecules, materials, and cells, but not for understanding phenomena on a scale that would still be submicroscopic if atoms were the size of marbles. Nanotechnology can't help here directly, but it can provide manufacturing facilities that will make huge particle accelerators economical, where today they strain national budgets. More generally, nanotechnology will help science wherever precision and fine details are important. Science frequently proceeds by trying small variations in almost identical experiments, comparing the results. This will be easier when molecular manufacturing can make two objects that are identical, molecule by molecule. In some areas, today's techniques are not only crude, but destructive. Archaeological sites are unique records of the human past, but today's techniques throw away most information during the dig, by accident. Future archaeologists, able to sift soil not speck by speck but molecule by molecule, will be grateful indeed to those archaeologists who today leave some ground undisturbed.

MEDICINE

Of all the areas where the ability to manufacture new tools is important, medicine is perhaps the greatest. The human body is intricate, and that intricacy extends beyond the range of human vision, beyond microscopic imaging, down to the molecular scale. "*Molecular medicine*" is an increasingly popular term today, but medicine today has only the simplest of molecular tools. As biology uses nanoinstruments to learn about disease and health, we will learn the physical requirements for restoring and maintaining health. And with this knowledge will come the tools needed to satisfy those requirements—tools ranging from improved pharmaceuticals to devices able to repair cells and tissues through *molecular surgery*.

Advanced medicine will be among the most complex and difficult applications of nanotechnology. It will require great knowledge, but nanoinstruments will help gather this knowledge. It will pose great engineering challenges, but computers of trillionfold greater power will help meet those challenges. It will solve medical problems on which we spend billions of dollars today, in hopes of modest improvements. Today, modern medicine often means an expensive way to prolong misery. Will nanomedicine be more of the same? Any reader over the age of, say, thirty knows how things start to go wrong: an ache here, a wrinkle there, the loss of an ability.

Over the decades, the physical quality of life declines faster and faster—the limits of what the body can do become stricter—until the limits are those of a hospital bed. The healing abilities we have when young seem to fade away. Modern medical practice expends the bulk of its effort on such things as intensive care units, dragging out the last few years of life without restoring health. Truly advanced medicine will be able to restore and supplement the youthful ability to heal. Its cost will depend on the cost of producing things more intricate than any we have seen before, the cost of producing computers, sensors, and the like by the trillions. To understand the prospects for medicine, like those for science and industry, we need to take a closer look at the cost of molecular manufacturing.

3

Molecular World

Nanotechnology is a bottom-up technology, building upward from the molecular scale. It will bring a revolution in human abilities like that brought by agriculture or power machinery. It can even be used to reverse many of the changes brought by agriculture or power machinery. But we humans are huge creations with no direct experience of the molecular world, and this can make nanotechnology hard to visualize, hence hard to understand. Scientists working with molecules face this problem today. They can often calculate how *molecules* will behave, but to *understand* this behaviour, they need more than heaps of numbers: they need pictures, movies, and interactive simulations, and so they are producing them at an ever-increasing pace. Recently a programm has been launched in "scientific visualization", in part to harness supercomputers to the problem of picturing the molecular world. Molecules are objects that exert forces on one another. If your hands were small enough, you could grab them, squeeze them, and bash them together. Understanding the molecular world is much like understanding any other physical world: it is a matter of understanding size, shape, strength, force, motion, and the like–a matter of understanding the differences between sand, water, and rock, or between steel and soap bubbles. Today's visualization tools give a taste of what will become possible with tomorrow's faster computers and better "virtual realities," simulated environments that let you tour a world that "exists" only as a model inside the computer.

Exploring the Molecular World

Joel Gregory manipulated molecules in the virtual reality of a simulated world using video goggles, tactile gloves, and a supercomputer. The early twenty-first century should be able to do even better. Imagine, then, that today you were to take a really long nap, oversleep, and wake up decades later in a nanotechnological world.

In the twenty-first century, even more than in the twentieth, it's easy to make things work without understanding them, but to a newcomer much of the technology seems like magic, which is dissatisfying. Back in the late twentieth century, most teaching used dry words and simple pictures, but now—for a topic like this—it's easier to explore a simulated world. And so you decide to explore a simulation of the molecular world.

Looking through the brochure, you read many tedious facts about the simulation: how accurate it is in describing sizes, forces, motions, and the like; how similar it is to working tools used by both engineering students and professionals; how you can buy one for your very own home, and so forth. It explains how you can tour the human body, see state-of-the-art nanotechnology in action, climb a bacterium, etc. For starters, you decide to take an introductory tour: simulations of real twentieth-century objects alongside quaint twentieth-century concepts of nanotechnology.

After paying a small fee and memorizing a few key phrases, you pull on a powersuit, pocket a Talking Tourguide, step into the simulation chamber, and strap the video goggles over your eyes. Looking through the goggles, you seem to be in a room with a table you know isn't really there and walls that seem too far away to fit in the simulation chamber. But trickery with a treadmill floor makes the walk to the walls seem far enough, and when you walk back and thump the table, it feels solid because the powersuit stops your hand sharply at just the right place. You can even feel the texture of the carvings on the table leg, because the suit's gloves press against your fingertips in the right patterns as you move. The simulation isn't perfect, but it's easy to ignore the defects. On the table is (or seems to be) an old 1990s silicon computer chip.

Vision and Motion

You feel as though you're falling toward the chip's surface, shrinking rapidly. In a moment, it looks roughly like Figure 1B, with your thumb still there holding it. The world grows blurrier, then everything seems to go wrong as you approach the molecular level. First, your vision blurs to uselessness—there is light, but it becomes a featureless fog. Your skin is tickled by small impacts, then battered by what feel like hard-thrown marbles. Your arms and legs feel as though they are

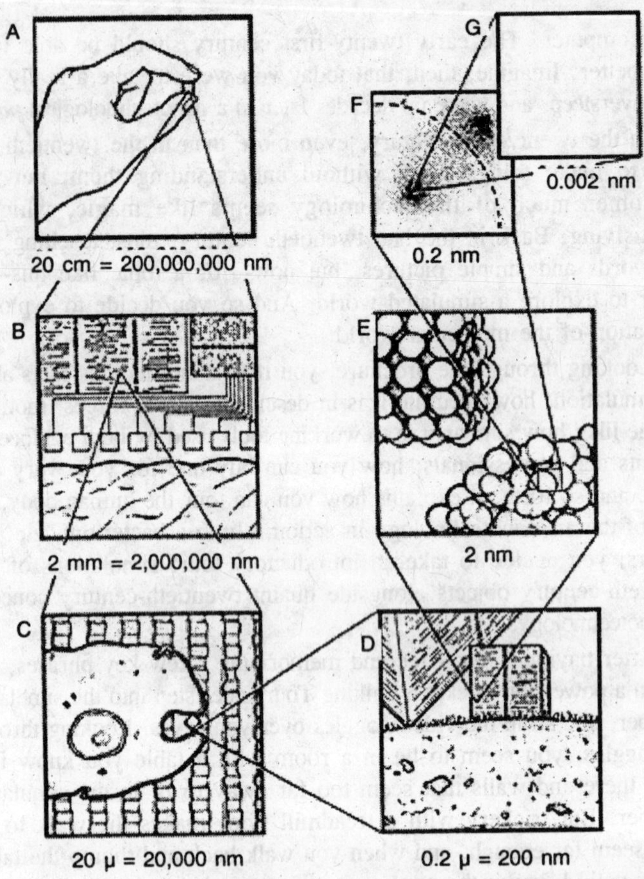

Fig. 3.1. Power of ten.

caught in turbulence, pulling to and fro, harder and harder. The ground hits your feet, you stumble and stick to the ground like a fly on flypaper, battered so hard that it almost hurts. You asked for realism, and only the built-in safety limits in the suit keep the simulated thermal motions of air molecules and of your own arms from beating you senseless. "Stop!" gives you a rest from the suit's yanking and thumping, and "Standard settings!" makes the world around you become more reasonable. The simulation changes, introducing the standard cheats. Your simulated eyes are now smaller than a light wave, making focus impossible, but the goggles snap your vision into sharpness and show the atoms around you as small spheres. (Real *nanomachines* are as blind as you were a moment ago, and can't cheat.) You are on the surface of the 1990s computer chip, between a cell and two blocky nanocomputers like the ones. Your simulated body is 50 nanometers tall, about 1/40,000,000 your real size, and the smaller nanocomputer is twice your height.

The simulation keeps bombarding you with air molecules, but the standard settings leave out the sensation of being pelted with marbles. A moment ago you were stuck tight to the ground by molecular stickiness, but the standard settings give your muscles the effective strength of steel—at least in simulation—by making everything around you much softer and weaker. The tourguide says that the only unreal features of the simulation have to do with you—not just your ability to see and to ignore thermal shaking and bombardment, but also your sheer existence at a size too small for anything so complex as a human being. It also explains why you can see things move, something about slowing down everything around you by a factor of 10 for every factor of 10 enlargement, and by another factor to allow for your being made stronger and hence faster. And so, with your greater strength and some adjustments to make your arms, legs, and torso less sticky, you can stand, see, feel, and take stock of the situation.

Molecular Texture

The ground underfoot, like everything around you, is pebbly with atom-sized bumps the size of your fingertips. Objects look like bunches of transparent grapes or fused marbles in a variety of pretty but imaginary colours. The simulation displays a view of atoms and molecules much like those used by chemists in the 1980s, but with a sharper 3-D image and a better way to move them and to feel the forces they exert. Actually, the whole simulation setup is nothing but an improved version of systems built in the late 1980s—the computer is faster, but it is calculating the same things. The video goggles are better and the whole-body powersuit is a major change, but even in the 1980s there were 3-D displays for molecules and crude devices that gave a sense of touching them.

The gloves on this suit give the sensation of touching whatever the computer simulates. When you run a fingertip over the side of the smaller nanocomputer, it feels odd, hard to describe. It is as if the surface were magnetic–it pulls on your fingertip if you move close enough. But the result isn't a sharp click of contact, because the surface isn't hard like a magnet, but strangely soft. Touching the surface is like touching a film of fog that grades smoothly into foam rubber, then hard rubber, then steel, all within the thickness of a sheet of corrugated cardboard. Moving sideways, your fingertip feels no texture, no friction, just smooth bumps more slippery than oil, and a tendency to get pulled into hollows. Pulling free of the surface takes a firm tug. The simulation makes your atom-sized fingertips feel the same forces that an atom would. It is strange how slippery the surface is—and it can't have been lubricated, since even a single oil molecule would be a lump the size of your thumb. This slipperiness makes it obvious how nanoscale bearings can work, how the parts of *molecular machines* can slide smoothly.

But on top of this, there is a tingling feeling in your fingers, like the sensation of touching a working loudspeaker. When you put your ear against the wall of the nanocomputer, you flinch back: for a moment, you heard a sound like the hiss of a twentieth—century television tuned to a channel with no broadcast, with nothing but snow and static—but loud, painfully loud. All the atoms in the surface are vibrating at high frequencies, too fast to see. This is thermal vibration, and it's obvious why it's also called thermal noise.

Gas and Liquid

Individual molecules still move too quickly to see. So, to add one more cheat to the simulation, you issue the command "Whoa!", and everything around seems to slow down by a factor of ten. On the surface, you now can see thermal vibrations that had been too quick to follow. All around, air molecules become easier to watch. They whiz about as thick as raindrops in a storm, but they are the size of marbles and bounce in all directions. They're also sticky in a magnetlike way, and some are skidding around on the wall of the nanocomputer. When you grab one, it slips away. Most are like two fused spheres, but you spot one that is perfectly round—it is an argon atom, and these are fairly rare. With a firm grip on all sides to keep it from shooting away like a watermelon seed, you pinch it between your steel-strong fingers. It compresses by about 10 percent before the resistance is more than you can overcome. It springs back perfectly and instantly when you relax, then bounces free of your grip. Atoms have an unfamiliar perfection about them, resilient and unchanging, and they surround you in thick swarms.

At the base of the wall is a churning blob that can only be a droplet of water. Scooping up a handful for a closer look yields a swarm of molecules, hundreds, all tumbling and bumbling over one another, but clinging in a coherent mass. As you watch, though, one breaks free of the liquid and flies off into the freer chaos of the surrounding air: the water is evaporating. Some slide up your arm and lodge in the armpit, but eventually skitter away. Getting rid of all the water molecules takes too much scraping, so you command "Clean me!" to dry off.

Too Small and Too Large

Beside you, the smaller *nanocomputer* is a block twice your height, but it's easy to climb up onto it as the tourguide suggests. Gravity is less important on a small scale: even a fly can defy gravity to walk on a ceiling, and an ant can lift what would be a truck to us. At a simulated size of fifty nanometers, gravity counts for nothing. Materials keep their strength, and are just as hard to bend or break, but the weight of an object becomes negligible. Even without the strength-enhancement that lets you overcome molecular stickiness, you could lift an object with 40 million times your mass–like a person of normal size lifting a box containing a half-dozen fully loaded oil tankers. To simulate this weak gravity, the powersuit cradles your body's weight, making you feel as if you were floating. This is almost like a vacation in an orbital theme park, walking with stickyboots on walls, ceilings, and whatnot, but with no need for antinausea medication.

On top of the nanocomputer is a stray protein molecule. This looks like a cluster of grapes and is about the same size. It even feels a bit like a bunch of grapes, soft and loose. The parts don't fly free like a gas or tumble and wander like a liquid, but they do quiver like gelatin and sometimes flop or twist. It is solid enough, but the folded structure is not as strong as your steel fingers. In the 1990s, people began to build molecular machinery out of proteins, copying biology. It worked, but it's easy to see why they moved on to better materials.

From a simulated pocket, you pull out a simulated magnifying glass and look at the simulated protein. This shows a pair of bonded atoms on the surface at 10 times magnification. The atoms are almost transparent, but even a close look doesn't reveal a nucleus inside, because it's too small to see. It would take 1,000 times magnification to be able to see it, even with the head start of being able to see atoms with your naked eye. How could people ever confuse big, plump atoms with tiny specks like nuclei? Remembering how your steel-strong fingers couldn't press more than a fraction of the way toward the nucleus of an argon atom from the air, it's clear why nuclear fusion is so difficult. In fact, the tourguide said that it would take a real-world projectile over a hundred times faster than a high-powered rifle bullet to penetrate into the atomic core and let two nuclei fuse.

Try as you might, there just isn't anything you could find in the molecular world that could reach into the middle of an atom to meddle with its nucleus. You can't touch it and you can't see it, so you stop squinting through the magnifying glass. Nuclei just aren't of much interest in nanotechnology.

Puzzle Chains

Taking the advice of the tourguide, you grab two molecular knobs on the protein and pull. It resists for a moment, but then a loop comes free, letting other loops flop around more, and the whole structure seems to melt into a writhing coil. After a bit of pulling and wrestling, the protein's structure becomes obvious: It is a long chain-longer than you are tall, if you could get it straight—and each segment of the chain has one of several kinds of knobs sticking off to the side. With the multicoloured, glassy-bead portrayal of atoms, the protein chain resembles a flamboyant necklace. This may be decorative, but how does it all go back together? The chain flops and twists and thrashes, and you pull and push and twist, but the original tight, solid packing is lost. There are more ways to go wrong in folding up the chain than there are in solving Rubik's Cube, and now that the folded structure is gone, it isn't even clear what the result should look like. How did those twentieth-century researchers ever solve the notorious "protein folding problem"? It's a matter of record that they started building protein objects in the late 1980s.

This protein molecule won't go back together, so you try to break it. A firm grip and a powerful yank straightens a section a bit, but the chain holds together and snaps back. Though unfolding it was easy, even muscles with the strength of steel—the strength of Superman—can't break the chain itself. *Chemical bonds* are amazingly strong, so it's time to cheat again. When you say, "Flimsy world-one second!" while pulling, your hands easily move apart, splitting the chain in two before its strength returns to normal. You've forced a chemical change, but there must be easier ways since chemists do their work without tiny superhands. While you compare the broken ends, they thrash around and bump together. The third time this happens, the chain rejoins, as strong as before. This is like having snap-together parts, but the snaps are far stronger than welded steel.

Modern *assembler* chemistry usually uses other approaches, but seeing this happen makes the idea of molecular assembly more understandable: Put the right pieces together in the right positions, and they snap together to make a bigger structure. Remembering the

"Whoa!" command, you decide to go back to the properly scaled speed for your size and strength. Saying "*Standard settings*!," you see the thrashing of the protein chain speed up to hard-to-follow blur.

Nanomachines

At your feet is a ribbed, ringed cylindrical object about the size of a soup can—not a messy, loosely folded strand like the protein, but a solid piece of modern nanotechnology. Picking it up, you can immediately feel how different it is from a protein. In the gear, *everything* is held in place by bonds as strong as those that strung together the beads of the protein chain. It can't unfold, and you'd have to cheat again to break its perfect symmetry. Like those in the wall of the nanocomputer, its solidly attached *atoms* vibrate only slightly. There's another gear nearby, so you fit them together and make the atomic teeth mesh, with bumps on one fitting into hollows on the other. They stick together, and the soft, slick atomic surfaces let them roll smoothly.

Underfoot is the nanocomputer itself, a huge mechanism built in the same rigid style. Climbing down from it, you can see through the transparent layers of the wall to watch the inner works. An electric motor an arm-span wide spins inside, turning a crank that drives a set of oscillating rods, which in turn drive smaller rods. This doesn't look like a computer; it looks more like an engineer's fantasy from the nineteenth century. But then, it is an antique design–the tourguide said that the original proposal was a piece of *exploratory engineering* dating from the mid-1980s, a mechanical design that was superseded by improved electronic designs before anyone had the tools to build even a prototype. This simulation is based on a version built by a hobbyist many years later.

The *mechanical nanocomputer* may be crude, but it does work, and it's a lot smaller and more efficient than the *electronic computers* of the early 1990s. It's even somewhat faster. The rods slide back and forth in a blur of motion, blocking and unblocking each other in changing patterns, weaving patterns of logic. This nanocomputer is a stripped-down model with almost no memory, useless by itself. Looking beyond it, you see the other block—which contains a machine powerful enough to compete with most computers built in 1990. This computer is a millionth of a meter on a side, but from where you stand, it looks like a blocky building looming over ten stories tall. The tourguide says that it contains over 100 billion atoms and stores as much data as a room full of books. You can see some of the storage system inside: row upon row of racks containing spools of molecular tape somewhat like the protein chain, but with simple bumps representing the 1s and 0s of computer data.

These nanocomputers seem big and crude, but the ground you're now standing on is also a computer–a single chip from 1990, roughly as powerful as the smaller, stripped-down nanocomputer at your side. As you gaze out over the chip, you get a better sense for just how crude things were a few decades ago. At your feet, on the smallest scale, the chip is an irregular mess. Although the wall of the nanocomputer is pebbly with atomic-scale bumps, the bumps are as regular as tile. The chip's surface, though, is a jumble of lumps and mounds. This pattern spreads for dozens of paces in all directions, ending in an irregular cliff marking the edge of a single transistor. Beyond, you can see other ridges and plateaus stretching off to the horizon. These form grand, regular patterns, the circuits of the computer. The horizon–the edge of the chip–is so distant that walking there from the center would (as the tourguide warns) take days. And these vast pieces of landscaping were considered twentieth-century miracles of miniaturization?

Cells and Bodies

Even back then, research in molecular biology had revealed the existence of smaller, more perfect machines such as the protein molecules in cells. A simulated human cell–put here because earlier visitors wanted to see the size comparisons–its on the chip next to the smaller nanocomputer. The tourguide points out that the simulation cheats a bit at this point, making the cell act as though it were in a watery environment instead of air. The cell dwarfs the nanocomputer, sprawling across the chip surface and rearing into the sky like a small mountain. Walking the nature trail around its edge would lead across many transistor-plateaus and take about an hour. A glance is enough to show how different it is from a nanocomputer or a gear: it looks organic, it bulges and curves like a blob of liver, but its surface is shaggy with waving molecular chains.

Walking up to its edge, you can see that the membrane wrapping the cell is fluid (cell *walls* are for stiff things like plants), and the membrane molecules are in constant motion. On an impulse, you thrust your arm through the membrane and poke around inside. You can feel many proteins bumping and tumbling around in the cell's interior fluid, and a crisscrossing network of protein cables and beams. Somewhere inside are the molecular machines that made all these proteins, but such bits of machinery are embedded in a roiling, organic mass. When you pull your arm out, the membrane flows closed behind. The fluid, dynamic structure of the cell is largely self healing. That's what let scientists perform experimental surgery on cells with the old, crude tools of the twentieth century: They didn't need to stitch up the holes they made when they poked around inside.

Even a single human cell is huge and complex. No real thinking being could be as small as you are in the simulation: A simple computer without any memory is twice your height, and the larger nanocomputer, the size of an apartment complex, is no smarter than one of the submoronic computers of 1990. Not even a bendable finger could be as small as your simulated fingers: in the simulation, your fingers are only one atom wide, leaving no room for the slimmest possible tendon, to say nothing of nerves.

For a last look at the organic world, you gaze out past the horizon and see the image of your own, full-sized thumb holding the chip on which you stand. The bulge of your thumb rises ten times higher than Mount Everest. Above, filling the sky, is a face looming like the Earth seen from orbit, gazing down. It is your own face, with cheeks the size of continents. The eyes are motionless. Thinking of the tourguide's data, you remember: The simulation uses the standard mechanical scaling rules, so being 40 million times smaller has made you 40 million times faster. To let you pull free of surfaces, it increased your strength by more than a factor of 100, which increased your speed by more than a factor of 10. So one second in the ordinary world corresponds to over 400 million here in the simulation. It would take years to see that huge face in the sky complete a single eyeblink.

Enough. At the command "Get me out!", the molecular world vanishes, and your feeling of weight returns as the suit goes slack. You strip off the video goggles—and hugely, slowly, blink.

4

Molecular Manufacturing

The sizes, forces, and general nature of objects in the molecular world are well known and been studied. Building on this, we can get a better picture of where developments seem to be leading, a better picture of *molecular manufacturing* itself.

Silicon Valley Faire

The molecular world showed some products of molecular manufacturing, but didn't show how they were made. The molecular technologies have mostly been replaced—but how did this happen? The Silicon Valley Faire is advertised as "An authentic theme park capturing life, work, and play in the early Breakthrough years." Since "work" must include manufacturing, it seems worth a visit. A broad dome caps the park —"To fully capture the authentic sights, sounds, and smells of the era," the tourguide politely says. Inside, the clothes and hairstyles, the newspaper headlines, the bumper-to-bumper traffic, all look much as they did before your long nap. A light haze obscures the buildings on the far side of the dome, your eyes burn slightly, and the air smells truly authentic.

Pocket Libraries

The Nanofabricators, Inc., plant offers the main display of early *nanotechnology*. As you near the building, the tourguide mentions that this is indeed the original manufacturing plant, given landmark status over twenty years ago, then made the centerpiece of the Silicon Valley Faire ten years later, when ... With a few taps, you reset the pocket tourguide to speak up less often. In the Nanofabricator plant, there's a moment of hushed quiet, a sense of walking into history. Nano-fabricators: home of the SuperChip, the first mass-market product of nanotechnology. It was the huge memory capacity of SuperChips that made possible the first Pocket Library. This section of the plant now houses a series of displays, including working replicas of early products. Picking up a Pocket Library, you find that it's not only the size of a wallet, but about the same weight. Yet it has enough memory to record every volume in the Library of Congress–something like a million times the capacity of a personal computer from 1990. It opens with a flip, the two-panel screen lights up, and a world of written knowledge is at your fingertips. Impressive.

Nanofabrication

In the workroom where SuperChips were first made, early nanotech manufacturing is spread out on display. The whole setup is surprisingly quiet and ordinary. Back in the 1980s and 1990s, chip plants had carefully controlled clean rooms with gowns and masks on workers and visitors, special workstations, and carefully crafted air flows to keep dust away from products. This room has none of that. It's even a little grubby.

In the middle of a big square table are a half-dozen steel tanks, about the size and shape of old-fashioned milk cans. Each can has a different label identifying its contents: *memory blocks*, *Data-transmission blocks*, *Interface blocks*. These are the parts needed for building up the chip. Clear plastic tubes, carrying clear and tea-coloured liquids, emerge from the mouths of the milk cans and drape across the table. The tubes end in fist-sized boxes mounted above shallow dishes sitting in a ring around the cans. As the different liquids drip into each dish, a beater like a kitchen mixer swirls the liquid. In each dish, *nanomachines* are building SuperChips.

A Nanofab "engineer," dressed in period clothing complete with name badge, is setting up a dish to begin building a new chip. "This," he says, holding up a blank with a pair of tweezers, "is a silicon chip like the ones made with pre-breakthrough technology. Companies here in this valley made chips like these by melting silicon, freezing it into lumps, sawing the lumps into slices, polishing the slices, and then going through a long series of chemical and photographic steps. When they were done, they had a pattern of lines and blobs of different materials on the surface. Even the smallest of these blobs contained *billions* of *atoms*, and it took several blobs working together to store a single bit of information. A chip this size, the size of your fingernail, could store only a fraction of a billion bits. Here at Nanofab, we used bare silicon chips as a base for building up nanomemory. The picture on the wall here shows the surface of a blank chip: no transistors, no memory circuits, just fine wires to connect up with the nanomemory we built on top. The nanomemory, even in the early days, stored *thousands* of billions of bits. And we made them like this, but a thousand at a time--" He places the chip in the dish, presses a button, and the dish begins to fill with liquid.

"A few years latter," he adds, "we got rid of the silicon chips entirely"-he props up a sign saying *this chip build began at*: 2:15 P.M., *estimated completion time*: 1:00 A.M.-" and we sped up the construction process by a factor of a thousand."

The chips in the dishes all look pretty much the same except for colour. The new chip looks like dull metal. The only difference you can see in the older chips, further along in the process, is a smooth rectangular patch covered by a film of darker material. An animated flowchart on the wall shows how layer upon layer of nanomemory building blocks are grabbed from solution and laid down on the surface to make that film. The tourguide explains that the energy for this process, like the energy for *molecular machines* within *cells*, comes from dissolved chemicals-from oxygen and fuel *molecules*. The total amount of energy needed here is trivial, because the amount of product is trivial: at the end of the process, the total thickness of nanomemory structure-the memory store for a Pocket Library-amounts to one-tenth the thickness of a sheet of paper, spread over an area smaller than a postage stamp.

Molecular Assembly

The animated flowchart showed nanomemory building blocks as big things containing about a hundred thousand atoms apiece (it takes a moment to remember that this is still submicroscopic). The build process in the dishes stacked these blocks to make the memory film on the SuperChip, but how were the blocks themselves built? The hard part in this molecular-manufacturing business has got to be at the bottom of the whole process, at the stage where molecules are put together to make large, complex parts.

The Silicon Valley Faire offers simulations of this molecular assembly process, and at no extra charge. From the tourguide, you learn that modern assembly processes are complex; that earlier processes-like those used by Nanofabricators, Inc. -used clever-but-obscure engineering tricks; and that the simplest, earliest concepts were never built. Why not begin at the beginning? A short walk takes you to the Museum of Antique Concepts, the first wing of the Museum of Molecular Manufacturing. A peek inside the first hall shows several people strolling around wearing loosely fitting jumpsuits with attached goggles and gloves, staring at nothing and playing mime with invisible objects. Oh well, why not join the fools' parade?

Stepping through the doorway while wearing the suit is entirely different. The goggles show a normal world outside the door and a molecular world inside. Now you, too, can see and feel the exhibit that fills the hall. It's much like the earlier simulated molecular world: it shares the standard settings for size, strength, and speed. Again, atoms seem 40 million times larger, about the size of your fingertips. This simulation is a bit less thorough than the last was--you can feel simulated objects, but only with your gloved hands. Again, everything seems to be made of quivering masses of fused marbles, each an atom.

Exploratory Engineering

Machinery fills the hall. Overall, the sight is reminiscent of an automated factory of the 1980s or 1990s. It seems clear enough what must be going on: Big machines stand beside a conveyor belt loaded with half-finished-looking blocks of some material; the machines must do some sort of work on the blocks. Judging by the conveyor belt, the blocks eventually move from one arm to the next until they turn a corner and enter the next hall.

Since nothing is real, the exhibit can't be damaged, so you walk up to a machine and give it a poke. It seems as solid as the wall of the nanocomputer in the previous tour. Suddenly, you notice something odd: no bombarding air molecules and no droplets of water–in fact, no loose molecules anywhere. Every atom seems to be part of a mechanical system, quivering thermal vibration, but otherwise perfectly controlled. Everything here is like the nanocomputer or like the tough little gear; none of it resembles the loosely coiled protein or the roiling mass of the living cell.

The conveyor belt seems motionless. At regular intervals along the belt are blocks of material under construction: workpieces. The nearest block is about a hundred marble-bumps wide, so it must contain something like 100 × 100 × 100 atoms, a full million. This block looks strangely familiar, with its rods, crank, and the rest. It's a nanocomputer–or rather, a blocklike part of a nanocomputer still under construction.

Standing alongside the pieces of nanocomputer on the conveyor belt, dominating the hall, is a row of huge mechanisms. Their trunks

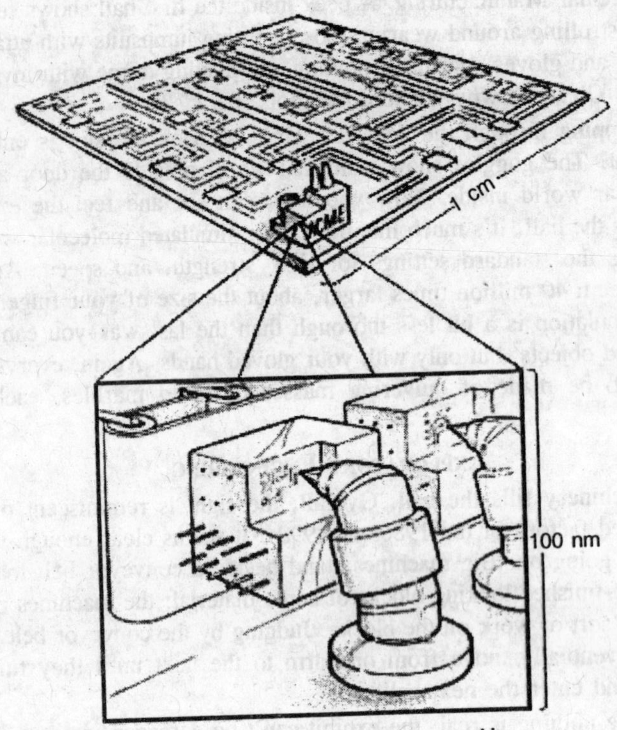

Fig. 4.1. Assembler with factory on chip.

rise from the floor, as thick as old oaks. Even though they bend over, they rear overhead. "Each machine," your tourguide says, "is the arm of a general-purpose molecular assembler.

One assembler arm is bent over with its tip pressed to a block on the conveyor belt. Walking closer, you see molecular assembly in action. The arm ends in a fist-sized knob with a few protruding marbles, like knuckles. Right now, two quivering marbles–atoms–are pressed into a small hollow in the block. As you watch, the two spheres shift, snapping into place in the block with a quick twitch of motion: a chemical reaction. The assembler arm just stands there, nearly motionless. The fist has lost two knuckles, and the block of nanocomputer is two atoms larger.

The tourguide holds forth: "This general-purpose assembler concept resembles, in essence, the factory robots of the 1980s. It is a computer-controlled mechanical arm that moves molecular tools according to a series of instructions. Each tool is like a single-shot stapler or rivet gun. It has a handle for the assembler to grab and comes loaded with a little bit of matter–a few atoms–which it attaches to the workpiece by a chemical reaction." This is like the rejoining of the protein chain in the earlier tour.

Molecular Precision

The atoms seemed to jump into place easily enough; can they jump out of place just as easily? By now the assembler arm has crept back from the surface, leaving a small gap, so you can reach in and poke at the newly added atoms. Poking and prying do no good: When you push as hard as you can (with your simulated fingers as strong as steel), the atoms don't budge by a visible amount. Strong molecular bonds hold them in place.

Your pocket tourguide–which has been applying the power of a thousand 1990s supercomputers to the task of deciding when to speak up–remarks, "Molecular bonds hold things together. In strong, stable materials atoms are either bonded, or they aren't, with no possibilities in between. Assemblers work by making and breaking bonds, so each step either succeeds perfectly or fails completely. In pre-breakthrough manufacturing, parts were always made and put together with small inaccuracies. These could add up to wreck product quality. At the molecular scale, these problems vanish. Since each step is perfectly precise, little errors can't add up. The process either works, or it doesn't."

But what about those definite, complete failures? Fired by scientific curiosity, you walk to the next assembler, grab the tip, and shake it. Almost nothing happens. When you shove as hard as you can, the tip moves by about one-tenth of an atomic diameter, then springs back. "Thermal vibrations can cause mistakes by causing parts to come together and form bonds in the wrong place," the tourguide remarks. "Thermal vibrations make floppy objects bend further than stiff ones, and so these assembler arms were designed to be thick and stubby to make them very stiff. Error rates can be kept to one in a trillion, and so small products can be perfectly regular and perfectly identical. Large products can be almost perfect, having just a few atoms out of place." This should mean high reliability. Oddly, most of the things you've been seeing outside have looked pretty ordinary–not slick, shiny, and perfect, but rough and homey. They must have been *manufactured* that way, or made by hand. Slick, shiny things must not impress anyone anymore.

Molecular Robotics

By now, the assembler arm has moved by several atom-widths. Through the translucent sides of the arm you can see that the arm is full of mechanisms: twirling shafts, gears, and large, slowly turning rings that drive the rotation and extension of joints along the trunk. The whole system is a huge, articulated robot arm. The arm is big because the smallest parts are the size of marbles, and the machinery inside that makes it move and bend has many, many parts. Inside, another mechanism is at work: The arm now ends in a hole, and you can see the old, spent molecular tool being retracted through a tube down the middle. Within a few minutes, a new tool is on its way back up the tube. Eventually, it reaches the end. Shafts twirl, gears turn, and clamps lock the tool in position. Other shafts twirl, and the arm slowly leans up against the workpiece again at a new site. Finally, with a twitch of motion, more atoms jump across, and the block is again just a little bit bigger. The cycle begins again. This huge arm seems amazingly slow, but the standard simulation settings have shifted speeds by a factor of over 400 million. A few minutes of simulation time correspond to less than a millionth of a second of real time, so this stiff, sluggish arm is completing about a million operations per second.

Peering down at the very base of the assembler arm, you can get a glimpse of yet more assembler-arm machinery underneath the floor: Electric motors spin, and a nanocomputer chugs away, rods pumping furiously. All these rods and gears move quickly, sliding and turning many times for every cycle of the ponderous arm. This seems inefficient; the mechanical vibrations must generate a lot of heat, so the electric motors must draw a lot of power. Having a computer control each arm is a lot more awkward now than it was in pre-breakthrough years. Back then, a robot arm was big and expensive and a computer was a cheap chip; now the computer is bigger than the arm. There must be a better way–but then, this is the Museum of Antique Concepts.

Building-Blocks into Buildings

Where do the blocks go, once the assemblers have finished with them? Following the conveyor belt past a dozen arms, you stroll to the end of the hall, turn the corner, and find yourself on a balcony overlooking a vaster hall beyond. Here, just off the conveyor belt, a block sits in a complex fixture. Its parts are moving, and an enormous arm looms over it like a construction crane. After a moment, the tourguide speaks up and confirms your suspicion: "After manufacturing, each block is tested. Large arms pick up properly made blocks. In this hall, the larger arms assemble almost a thousand blocks of various kinds to make a complete nanocomputer.

The grand hall has its own conveyor belt, bearing a series of partially completed nanocomputers. Arrayed along this grand belt is a row of grand arms, able to swing to and fro, to reach down to lesser conveyor belts, pluck million-atom blocks from testing stations, and plug them into the grand workpieces, the nanocomputers under construction. The belt runs the length of the hall, and at the end, finished nanocomputers turn a corner–to a yet-grander hall beyond?

After gazing at the final-assembly hall for several minutes, you notice that nothing seems to have moved. Mere patience won't do: at the rate the smaller arms moved in the hall behind you, each block must take months to complete, and the grand block-handling arms are taking full advantage of the leisure this provides. Building a computer, start to finish, might take a terribly long time. Perhaps as long as the blink of an eye.

Molecular assemblers build blocks that go to block assemblers. The block assemblers build computers, which go to system assemblers, which build systems, which–at least one path from molecules to large products seems clear enough. If a car were assembled by normal-sized robots from a thousand pieces, each piece having been assembled by smaller robots from a thousand smaller pieces, and so on, down and down, then only ten levels of assembly process would separate cars from molecules. Perhaps, around a few more corners and down a few more ever-larger halls, you would see a post-breakthrough car in the making, with unrecognizable engine parts and comfortable seating being snapped together in a century-long process in a hall so vast that the Pacific Ocean would be a puddle in the corner... Just ten steps in size; eight, starting with blocks as big as the ones made in the hall behind you. The molecular world seems closer, viewed this way.

Molecular Processing

Stepping back into that hall, you wonder how the process begins. In every cycle of their sluggish motion, each molecular assembler gets a fresh tool through a tube from somewhere beneath the floor, and that somewhere is where the story of molecular precision begins. And so you ask, "Where do the tools come from?", and the tourguide replies, "You might want to take the elevator to your left." Stepping out of the elevator and into the basement, you see a wide hall full of small conveyor belts and pulleys; a large pipe runs down the middle. A plaque on the wall says, "Mechanochemical processing concept, circa 1990." As usual, all the motions seem rather slow, but in this hall everything that seems designed to move is visibly in motion. The general flow seems to be away from the pipe, through several steps, and then up through the ceiling toward the hall of assemblers above.

After walking over to the pipe, you can see that it is nearly transparent. Inside is a seething chaos of small molecules: the wall of the pipe is the boundary between loose molecules and controlled ones, but the loose molecules are well confined. In this simulation, your fingertips are like small molecules. No matter how hard you push, there's no way to drive your finger through the wall of the pipe. Every few paces along the pipe a fitting juts out, a housing with a mechanically driven rotating thing, exposed to the liquid inside the pipe, but also exposed to a belt running over one of the pulleys, embedded in the housing. It's hard to see exactly what is happening.

The tourguide speaks up, saying, "Pockets on the rotor capture single molecules from the liquid in the pipe. Each rotor pocket has a size and shape that fits just one of the several different kinds of molecule in the liquid, so the process is rather selective. Captured molecules are then pushed into the pockets on the belt that's wrapped over the pulley there, then–" "Enough," you say. Fine, it singles out molecules and sticks them into this maze of machinery. Presumably, the machines can sort the molecules to make sure the right kinds go to the right places.

The belts loop back and forth carrying big, knobby masses of molecules. Many of the pulleys–rollers?–press two belts together inside a housing with auxiliary rollers. While you are looking at one of these, the tourguide says, "Each knob on a belt is a mechanochemical-processing device. When two knobs on different belts are pressed together in the right way, they are designed to transfer molecular fragments from one to another by means of a mechanically forced chemical reaction. In this way, small molecules are broken down, recombined, and finally joined to molecular tools of the sort used in the assemblers in the hall above. In this device here, the rollers create a pressure equal to the pressure found halfway to the center of the Earth, speeding a reaction that–"

"Fine, fine," you say. Chemists in the old days managed to make amazingly complex molecules just by mixing different chemicals together in solution in the right order under the right conditions. Here, molecules can certainly be brought together in the right order, and the conditions are much better controlled. It stands to reason that this carefully designed maze of pulleys and belts can do a better job of molecule processing than a test tube full of disorganized liquid ever could. From a liquid, through a sorter, into a mill, and out as tools: this seems to be the story of molecule processing. All the belts are loops, so the machinery just goes around and around, carrying and transforming molecular parts.

BEYOND ANTIQUES

This system of belts seems terribly simple and efficient, compared to the ponderous arms driven by frantic computers in the hall above. Why stop with making simple tools? You must have muttered this, because the tourguide speaks up again and says, "The Special-Assembler Exhibit shows another early molecular-manufacturing concept that uses the principles of this molecule-processing system to build large, complex objects. If a system is building only a single product, there is no need to have computers and flexible arms move parts around. It is far more efficient to build a machine in which everything just moves on belts at a constant speed, adding small parts to larger ones and then bringing the larger ones together as you saw at the end of the hall above."

This does seem like a more sensible way to churn out a lot of identical products, but it sounds like just more of the same. Gears like fused marbles, belts like coarse beadwork, drive shafts, pulleys, machines and more machines. In a few places, marbles snap into new patterns to prepare a tool or make a product. Roll, roll, chug, chug, pop, snap, then roll and chug some more. As you leave the simulation hall, you ask, "Is there anything important I've missed in this molecular manufacturing tour?"

The tourguide launches into a list: "Yes–the inner workings of assembler arms, with drive shafts, worm gears, and harmonic drives; the use of Diels-Alder reactions, interfacial free-radial chain reactions, and dative-bond formation to join blocks together in the larger-scale stages of assembly; different kinds of mechanochemical processing for preparing reactive molecular tools; the use of staged-cascade methods in providing feed-molecules of the right kinds with near-perfect reliability; the differences between efficient and inefficient steps in molecular processing; the use of redundancy to ensure reliability in large systems despite sporadic damage; modern methods of building large objects from smaller blocks; modern electronic nanocomputers; modern methods for–"

Crude Technology

Molecular manufacturing works much like ordinary manufacturing, but with devices built so small that a single loose molecule of pollutant would be like a brick heaved into a machine tool. John Walker of

Autodesk, a leading company in computer-aided design, observes that nanotechnology and today's crude methods are very different: "Technology has never had this kind of precise control; all of our technologies today are *bulk technologies*. We take a big chunk of stuff and hack away at it until we're left with the object we want, or we assemble parts from components without regard to structure at the molecular level."

Molecular manufacturing will orchestrate atoms into products of symphonic complexity, but modern manufacturing mostly makes loud noises. These figurative noises are sometimes all too literal: A crack in a metal forging grows under stress, a wing fails, and a passenger jet crashes from the sky. A chemical reaction goes out of control, heat and pressure build, and a poisonous blast shakes the countryside. A lifesaving product cannot be made, a heart fails, and a hospital's heart-monitoring machine signals the end with a high-pitched wail.

At present many things are made from metal, by machining. From the perspective of our standard, simulated molecular world, a typical metal part is a piece of terrain many days' journey across. The metal itself is weak compared to the bonds of the protein chain or other tough nanomechanisms: solid steel is no stronger than your simulated fingers, and the atoms on its surface can be pushed around with your bare hands. Standing on a piece of metal being machined in a lathe, you would see a cutting blade crawl past a few times per year, like a majestic plough the size of a mountain range. Each pass would rip up a strip of the metal landscape, leaving a rugged valley broad enough to hold a town. This is machining from a nanotechnological perspective: a process that hacks crude shapes from intrinsically weak materials.

Similarly, electronics are made from silicon chips. We have already seen the landscape of a finished chip. During manufacturing, metal features would be built up by a centuries-long drizzle of metal-atom rain, and hollows would be formed by a centuries-long submergence in an acid sea. From the perspective of our simulation, the whole process would resemble geology as much as manufacturing, with the slow layering of sedimentary deposits alternating with ages of erosion. The term *nanotechnology* is sometimes used as a name for small-scale microtechnology, but the difference between molecular manufacturing and this sort of microlandscaping is like the difference between watchmaking and bulldozing.

Now-a-days, chemists make molecules by solution chemistry. We have seen what a liquid looks like in our first simulation, with molecules bumping and tumbling and wandering around. Just as assemblers can make chemical reactions occur by bringing molecules together mechanically, so reactions can occur when molecules bump at random through thermal vibration and motion in a liquid. Indeed, much of what we know today about chemical reactions comes from observing this process. Chemists make large molecules by mixing small molecules in a liquid. By choosing the right molecules and conditions, they can get a surprising measure of control over the results: only some pairs of molecules will react, and then only in certain ways. Doing chemistry this way, though, is like trying to assemble a model car by putting the pieces in a box and shaking. This will only work with cleverly shaped pieces, and it is hard to make anything very complex. Chemists today consider it challenging to make a precise, three-dimensional structure having a hundred atoms, and making one with a thousand atoms is a great accomplishment. Molecular manufacturing, in contrast, will routinely assemble millions or billions. The basic chemical principles will be the same, but control and reliability will be vastly greater. It is the difference between throwing things together blindly and putting them together with a watchmaker's care.

Technology today doesn't permit thorough control of the structure of matter. Molecular manufacturing will. Today's technologies have given us computers, spacecraft, indoor plumbing, and the other wonders of the modern age. Tomorrow's will do much more, bringing change and choices.

Simple Matter, Smart Matter

Today's technology mostly works with matter in a few basic forms: gases, liquids, and solids. Though each form has many varieties, all are comparatively simple.

Gases, consist of molecules ricocheting through space. A volume of gas will push against its walls and, if not walled in, expand without limit. Gases can supply certain raw materials for nanomachines, and nanomachines can be used to remove pollutants from air and turn them into something else. Gases lack structure, so they will remain simple.

Liquids are somewhat like gases, but their molecules cling together to form a coherent blob that won't expand beyond a certain limit. Liquids will be good sources of raw materials for nanomachines because they are denser and can carry a wide range of fuels and raw materials in solution (the pipe in the molecular-processing hall contained liquid).

Nanomachines can clean up polluted water as easily as air, removing and transforming noxious molecules. Liquids have more structure than gases, but nanotechnology will have its greatest application to solids.

Solids are diverse. Solid butter consists of molecules stronger than steel, but the molecules cling to one another by the weaker forces of molecular stickiness. A little heat increases thermal vibrations and makes the solid structure disintegrate into a blob of liquid. Butterlike materials would make poor nanomachines. Metals consist of atoms held together by stronger forces, and so they can be structurally stronger and able to withstand higher temperatures. The forces are not very directional, though, and so planes of metal atoms can slip past one another under pressure; this is why spoons bend, rather than break. This ability to slip makes metals less brittle and easier to shape (with crude technology), but it also weakens them. Only the strongest, hardest, highest-melting-point metals are worth considering as parts of nanomachines.

Diamond consists of carbon atoms held together by strong, directional bonds, like the bonds down the axis of a protein chain. These directional bonds make it hard for planes of atoms to slip past one another, making diamond (and similar materials) very strong indeed–ten to a hundred times stronger than steel. But the planes can't easily slip, so when the material fails, it doesn't bend, it breaks. Tiny cracks can easily grow, making a large object seem weak. Glass is a similar material: glass windows seem weak–and a scratch makes glass far weaker–yet thin, perfect glass fibers are widely used to make composite materials stronger and lighter than steel. Nanotechnology will be able to build with diamond and similar strong materials, making small, flawless fibers and components.

In engineering today, diamond is just beginning to be used. Japan has pioneered a technology for making diamond at low pressure, and a Japanese company sells a speaker with excellent high-frequency response–the speaker cone is reinforced with a light, stiff film of diamond. Diamond is extraordinary stuff, made from cheap materials like natural gas. U.S. companies are scrambling to catch up.

All these materials are simple. More complex structures lead to more complex properties, and begin to give some hint of what molecular manufacturing will mean for materials.

What if you strung carbon atoms in long chains with side-groups, a bit like a protein chain, and linked them into a big three-dimensional mesh? If the chains were kinked so that they couldn't pack tightly,

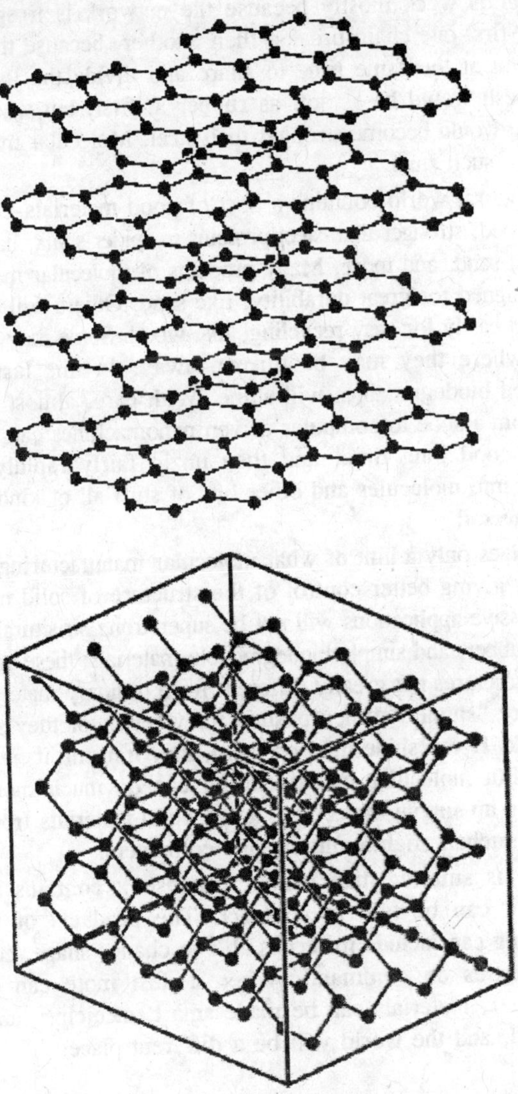

Fig. 4.2. Carbon-soft and hard.

they would coil up and flop around almost like molecules in a liquid, yet the strong bonds would keep the overall mesh intact. Pulling the whole network would tend to straighten the chains, but their writhing motions would tend to coil them back up. This sort of network has been made: it is called rubber.

Rubber is weak mostly because the network is irregular. When stretched, first one chain breaks, then another, because they don't all become taut at the same time to share and divide the load. A more regular mesh would be as soft as rubber at first, but when stretched to the limit would become stronger than steel. Molecular manufacturing could make such stuff.

The natural world contains a host of good materials–*cellulose* and *lignin* in wood, stronger-than-steel proteins in spider's silk, hard ceramics in grains of sand, and more. Many products of molecular manufacturing will be designed for great durability, like sand. Others will be designed to fall apart easily for easy recycling, like wood. Some may be designed for uses where they may be thrown away. In this last category, nanotailored biodegradables will shine. With care, almost any sort of product from a shoe to computer-driven nanomachines can be made to last for a good long time, and then unzip fairly rapidly and very thoroughly into molecules and other bits of stuff all of kinds normally found in the soil.

This gives only a hint of what molecular manufacturing will make possible by giving better control of the structure of solid matter. The most impressive applications will not be superstrong structural materials, improved rubber, and simple biodegradable materials: these are uniform, repetitive structures not greatly different from ordinary materials. These materials are "stupid." When pushed, they resist, or they stretch and bounce back. If you shine light on them, they transmit it, reflect it, or absorb it. But molecular manufacturing can do much more. Rather than heaping up simple molecules, it can build materials from trillions of motors, ratchets, light-emitters, and computers.

Muscle is smarter than rubber because it contains molecular machines: it can be told to contract. The products of molecular manufacturing can include materials able to change shape, colour, and other properties on command. When a dust mote can contain a supercomputer, materials can be made smart, medicine can be made sophisticated, and the world will be a different place.

Origins

The idea of *molecular nanotechnology*, like most ideas, has roots stretching far back in time. In ancient Greece, Democritus suggested that the world was built of durable, invisible particles–atoms, the building blocks of solid objects, liquids, and gases. In the last hundred years, scientists have learned more and more about these building blocks, and chemists have learned more and more ways to combine them to make new things. Decades ago, biologists found molecules that do complex things; they termed them "molecular machines."

Physicist *Richard Feynman* was a visionary of miniaturization who pointed toward something like molecular nanotechnology: on December 1959, proposed that large machines could be used to make smaller machines, which could make still smaller ones, working in a top-down fashion from the macroscale to the microscale. At the end of his talk, he painted a vision of moving individual atoms, pointing out, "The principles of physics, as far as I can see, do not speak against the possibility of maneuvering things atom by atom." He pictured making molecules, pointing clearly in the direction taken by the modern concept of nanotechnology: "But it is interesting that it would be, in principle, possible (I think) for a physicist to synthesize any chemical substance the chemist writes down. Give the orders, and the physicist synthesizes it. How? Put the atoms down where the chemist says, and so you make the substance."

Despite this clear signpost pointing to a potentially revolutionary area, no one filled the conceptual gap between miniature machines and chemical substances. There was no clear concept of making molecular machines able to build more such machines, no notion of controllable molecular manufacturing. With hindsight, one wonders why the gap took so long to fill. Feynman himself didn't follow it up, saying that the ability to maneuver atoms one by one "will really be useless" since chemists would come up with traditional, bulk-process ways to make new chemical substances. For a researcher whose main interest was physics, he had contributed much just by placing the signpost: it was up to others to move forward. Instead, the idea of molecular machines for molecular manufacturing didn't appear for decades.

From today's viewpoint, molecular nanotechnology looks more like an extension of chemistry than like an extension of miniaturization. A mechanical engineer, looking at nanotechnology, might ask, "How can machines be made so small?" A chemist, though, would ask, "How can molecules be made so large?" The chemist has the better question. Nanotechnology isn't primarily about miniaturizing machines, but about extending precise control of molecular structure to larger and larger scales. Nanotechnology is about making (precise) things big.

Nature gives the most obvious clues to how this can be done, and it was the growing scientific literature on natural molecular machines that led one of the present authors (Drexler) to propose *molecular*

Table 4.1. Macroscopic and Molecular Components

Technology	*Function*	*Molecular Examples*
Struts, beams, casins	Transmit force, hold positions	Cell walls, microtubules
Cables	Transmit tension	Collagen, silk
Fasteners, glue	Connect parts	Intermolecular forces
Solenoids, actuators	Move things	Muscle actin, myosin
Motors	Turn shafts	Flagellar motor
Drive shafts	Transmit torque	Bacterial flagella
Bearings	Support moving parts	Single bonds
Clamps	Hold workpieces	Enzymatic binding sites
Tools	Modify workpieces	Enzymes, reactive molecules
Production lines	Control devices	Enzyme systems, ribosomes
Numerical control systems	Store and read programs	Genetic system

nanotechnology of the sort described here. A strategy to reach the goal was part of the concept: Build increasingly complex molecular machinery from simpler pieces, including molecular machines able to build more molecular machines. And the motivation for studying this, and publishing? Largely the fear of living in a world that might rush into the new technology blindly, with ugly consequences. This concept and initial exploratory work started in early 1977 at MIT; the first technical publication came in 1981 in the *Proceedings of the National Academy of Sciences*. For years, MIT remained the center of thinking on nanotechnology and molecular manufacturing: in 1985, the MIT Nanotechnology Study Group was formed; it soon initiated an annual lecture series which grew into a two-day symposium by 1990.

The first book on the topic, *Engines of Creation*, was published in 1986. In 1988, Stanford University became the first to offer a course in molecular nanotechnology, sponsored by the Department of Computer Science. In 1989, this department hosted the *first major conference* on the subject, cosponsored by the Foresight Institute and Global Business Network. With the upcoming publication of a *technical book describing nanotechnology*—from molecular mechanical and quantum-mechanical principles up to assembly systems and products–the subject will be easier to teach, and more college courses will become available.

In parallel with the development and spread of ideas about nanotechnology and molecular manufacturing—ideas that remain pure theory, however well grounded–scientists and engineers, working in laboratories to build real tools and capabilities, have been pioneering roads to nanotechnology. Research has come a long way since the mid-1980s, as we'll see in the next chapter. But, as one might expect with a complex new idea that, if true, disrupts a lot of existing plans and expectations, some objections have been heard.

"It won't work"

Life might be much simpler if these ideas about nanotechnology had some fatal flaw. If only molecules couldn't be used to form machines, or the machines couldn't be used to build things, then we might be able to keep right on going with our crude technologies: our medicine that doesn't heal, our spacecraft that don't open a new frontier, our oil crises, our pollution, and all the limits that keep us from trading familiar problems for strange ones. Most new ideas are wrong, especially if they purport to bring radical changes. It is not unreasonable to hope that these are wrong. From years of discussions with chemists, physicists, and engineers, it is possible to compile what seems to be a complete list of basic, critical questions about whether nanotechnology will work. The questioners generally seem satisfied with the answers.

"Will thermal vibrations mess things up?"

The earlier scenarios describe the nature of thermal vibration and the problems it can cause. Designing nanomachines strong enough and stiff enough to operate reliably despite thermal vibration is a genuine engineering challenge. But calculating the design requirements usually requires only simple textbook principles, and these requirements can be met for everything described in this book.

"Will quantum uncertainty mess things up?"

Quantum mechanics says that particles must be described as small smears of probability, not as points with perfectly defined locations. This is, in fact, why the atoms and molecules in the simulations felt so soft and smooth: their electrons are smeared out over the whole volume of the molecule, and these electron clouds taper off smoothly and softly toward the edges. Atoms themselves are a bit uncertain in position, but this is a small effect compared to thermal vibrations. Again, simple textbook principles apply, and well-designed molecular machines will work.

"Will loose molecules mess things up?"

Chemists work with loose molecules in liquids, and they naturally tend to picture molecules as flying around loose. It is possible to build nanomachines and molecular-manufacturing systems that work in this sort of environment (biological mechanisms are an existence proof), but in the long run, there will be no need to do so. The Silicon Valley Faire simulation gives the right idea: Systems can be built with no loose molecules, making nanomechanical design much easier. If no molecules are loose inside a machine, then loose molecules can't cause problems there.

"Will chemical instability mess things up?"

Chemists perform chemical reactions, which means that they tend to work with reactive, unstable molecules. Many molecules, though, can sit around in peace with their neighbours for millions of years, as is known both from chemical theory and from the study of molecules trapped in ancient rock.

Nanomachines can be built from the more stable sorts of structure. The only necessary exception is in molecular assembly, where molecules must react, but even here the reactive molecules need not be turned loose. They can be applied just when and where they are needed in the construction process.

"Is it too complex, like biology?"

An easy way to explain molecular manufacturing is to say that it is somewhat like molecular biology: small, complex molecular devices working together to build things and do various jobs. The next point, however, is that molecular manufacturing is different in every detail and different in overall structure: compare the nanocomputers, assembler arms, and conveyor belts described above to the shaggy, seething living cell described in the last chapter. Biology is complex in a strange and wonderful way. Engineers need not even understand life, much less duplicate it, merely to build a molecular-scale factory.

Nanotechnology is basically a shotgun marriage of chemistry and mechanical engineering, with physics (as always) presiding. This makes a complete evaluation difficult for most of today's specialists, because each of these fields is taught separately and usually practiced separately. Many specialists, having highly focused backgrounds, find themselves unequipped to evaluate proposals that overlap other disciplines. When asked to do so, they will state feelings of discomfort, because although they can't identify any particular problems, they can't verify the entire concept as sound. Scientists and engineers with multidisciplinary backgrounds, or with access to specialists from other fields, can evaluate the idea from all sides.

It will work.

When physicists, chemists, biologists, engineers, and computer scientists evaluate those parts of nanotechnology that fall within their disciplines, they agree: At no point would it require new principles or violate a physical law. There may for many years be some experts offering negative off-the-cuff opinions, but the consensus among those who have taken the time to examine the facts is clear. Molecular nanotechnology falls entirely within the realm of the possible.

"It would work, but isn't it a bad idea to implement it?"

If this means, "These new technologies could easily do far more harm than good," then there is no argument, because no one seems to disagree. If this means, "These new technologies will certainly do more harm than good," then we disagree: much good is possible, much harm is avoidable, and it would be too bold to declare any such outcome "certain." Increased human abilities have routinely been used to damage the environment and to make war. Even the crude technologies of the twentieth century have taken us to the brink. It is natural to feel exhilarated (or terrified) by a prospect that promises (or threatens) to extend human abilities beyond most past dreams (or nightmares). It is better to feel both, to meld and moderate these feelings, and to set out on a course of action that makes bad outcomes less likely. We're convinced that the best course is to focus on the potential good while warning of the potential evils.

Perspective

We are still many years away from nanotechnology based on molecular manufacturing. It might even seem that such vast, slow giants as ourselves could never make such small, quick machines. The following sections will describe how advances in science and technology are leading toward these abilities. We'll try to get some feel for the road ahead, for its length, and for how fast we're moving. We are already surprisingly close to developing a crude molecular manufacturing technology, and getting visibly closer every week. The first, crude technology will enable the construction of molecular machines that can be used to build better molecular machines, climbing a ladder of capabilities that leads to general-purpose molecular assemblers as good or better than those described here. The opportunities then will be enormous. If we haven't prepared, the dangers, too, will be enormous. Whether we're ready or not, the resulting changes will be disruptive, sweeping industries aside, upending military strategies, and transforming our ways of life.

5

NANOMOLECULES

Most contemporary industrial fabrication processes are based on "top-down" technologies, wherein small objects are sawn or machined from larger objects, or small features are imposed on larger objects, in either case by removing unwanted matter. The results of such processes may be small, such as micron-featured integrated circuits, or very large, such as jet aircraft, but in most cases the material is being processed in chunks far larger than molecular scale.

Molecular manufacturing, on the other hand, represents a "bottom-up" technology. Desired products will be built directly by "assembler" machines, molecule by molecule, making larger and larger objects with atomic precision. The results of such processes may also be very small or very large, much as biology builds both micron-sized bacteria and 100-meter tall sequoia trees. However, since assemblers add matter only where it is intended, little need be removed and hence there may be minimal waste during the process.

By guiding with precision the assembly of molecules and supramolecular structures, such a manufacturing system could construct an extraordinarily wide range of products of unprecedented quality and performance. In 1959, the Nobel physicist Richard P. Feynman observed that: "The principles of physics, as far as I can see, do not speak against the possibility of maneuvering things atom by atom. It is not an attempt to violate any laws; it is something, in principle, that can be done; but, in practice, it has not been done because we are too big ... Ultimately, we can do chemical synthesis." "A chemist comes to us and says, 'Look, I want a molecule that has the atoms arranged thus and so; make me that molecule.' The chemist does a mysterious thing when he wants to make a molecule. He sees that it has got that ring, so he mixes this and that, and he shakes it, and he fiddles around. And, at the end of a difficult process, he usually does succeed in synthesizing what he wants."

"But it is interesting that it would be, in principle, possible (I think) for a physicist to synthesize any chemical substance that the chemist writes down. Give the orders and the physicist synthesizes it. How? Put the atoms down where the chemist says, and so you make the substance. The problems of chemistry and biology can be greatly helped if our ability to see what we are doing, and to do things on an atomic level, is ultimately developed a development which I think cannot be avoided."

Nearly 40 years after Feynman's famous "Plenty of Room at the Bottom" speech, and a decade after Drexler's original proposal for a bottom-up approach to machine-building using molecular assemblers, Nobel chemist Richard Smalley also largely agreed that this objective should prove feasible. Noted Smalley: "On a length scale of more than one nanometer, the mechanical robot assembler metaphor envisioned by Drexler almost certainly will work..."

Many skeptical questions arise when one first encounters the ideas of molecular nanotechnology and molecular assemblers. It is useful to keep in mind the proven feasibility of such systems in the biological tradition over billions of years of natural evolution. In one sense, molecular nanotechnology will be a refinement and expansion upon how nature works at the molecular scale. Nature's examples, such as human beings, certainly answer the most basic skeptical questions, such as: Can macroscopic objects be built from molecular scale processes. Are molecular objects stable. Observes Nobel chemist Jean-Marie Lehn: "The chemist finds illustration, inspiration and stimulation in natural processes, as well as confidence and reassurance since they are proof that such highly complex systems can indeed be achieved on the basis of molecular components."

What about quantum effects? The uncertainty principle makes electron positions somewhat fuzzy, but the atom as a whole has a comparatively definite position set by the relatively great mass of the atomic nucleus. The quantum probability function of electrons in atoms tends to drop off exponentially with distance outside the atom, giving atoms a moderately sharp "edge". How about the effects of Brownian bombardment on nano-machines? Describing a nanomechanical component as a harmonic oscillator embedded in a gas, Drexler notes:

"At equilibrium, an impinging gas molecule is as likely to absorb energy as to deliver it, and so molecular bombardment has no net effect on the amplitude of vibration. How a system is coupled to a thermal bath can affect its detailed dynamics, but not the statistical distribution of dynamical quantities." Nanomachines must also obey the laws of thermodynamics. Nanorobots cannot be used as a "Maxwell's demon"— energy must be expended to do useful work; there can be no free lunch.

Will high-energy radiation damage nanomachines? Radiation can break chemical bonds and disrupt molecular machines. The annual failure rate for a properly designed ~1 micron molecular machine (containing many billions of atoms) can be made as low as several percent. This estimate is derived from a consideration of all known thermal, photochemical, radiation, and other damage mechanisms, but the dominant error mechanism that appears difficult to substantially reduce is damage caused by background radiation. Failure rates are not zero, but are nonetheless remarkable by today's standards, especially given error detection and correction at the module level.

During nanodevice fabrication, the problems of cross-bonding and reactive intermediates pose additional constraints that must be dealt with using appropriate assembler designs. Specifically, assemblers with rigid positioning components operating in vacuo can restrict reactive intermediates and molecular tool tips to desired locations with a precision of at least ~0.1 nm, or slightly less than an atomic diameter. Within the confines of an evacuated or "*eutactic*" assembler workspace, the location of every structural atom may be known to within the uncertainty created by thermal noise. Internal structural components may possess relatively high stiffness, so positional uncertainty caused by thermal noise may be small. By employing sufficiently stiff mechanisms the position of every structural atom in the system can be known to within a fraction of an atomic diameter with high reliability and without the need for explicit positional sensing. Steric tool hindrance —in molecular manufacturing, the conical volume occupied by a tool tip or manipulator platform—is a technical design problem that may be overcome by using sufficiently stiff tools or multi-tip tools, working in vacuo. Additionally, while some synthetic methods might involve the manipulation of individual atoms by appropriate tools, such as the hydrogen abstraction tool which removes a single selected hydrogen atom from a diamondoid surface, other reactions will involve small clusters of atoms or larger molecular components.

Medical nanorobots may be built to atomic precision, but once built and deployed such devices often will not perform molecularly precise movements during normal functions and may frequently rely upon sensor-based motion control.

What about friction and wear among nanomechanical components? Properly designed molecular machines lack wear mechanisms, although other damage mechanisms remain. Dry bearings with atomically precise surfaces can have negligible static friction and wear, and very low dynamic friction or drag. Nanomechanical components are better viewed as moving smoothly in a force field than as sliding subject to friction. For example, total drag power dissipated by an isolated molecular sleeve bearing of radius 2 nm spinning at 1 MHz is dominated by band-stiffness scattering, amounting to 0.000004 picowatts (pW) very small compared to the 1-1000 pW power draw that will be typical of micron-size medical nanorobots. The strong precise surfaces of nanomachines experience no change during a typical operational cycle, hence zero wear. Within the single-point failure model , the first step in a wear process (e.g., a dislocated atom) is regarded as fatal, hence cumulative wear plays no role in determining device lifetimes.

In eutactic nanomachines, contaminants of all kinds are rigorously excluded. The equivalent of wear particles cannot appear until the device has failed. On this scale, repulsive fields provide "lubrication" and an oil molecule would be a contaminant object, not a lubricant. The general conclusion is that molecular nanotechnology violates no physical laws. In 1998, it was generally accepted that molecular nanotechnology will be developed, although there remains some disagreement about how long it will take. To progress from today's limited capabilities to the complex nanorobots described later in this trilogy will clearly require a great deal of research and development effort. Still, the proper question is no longer "if" but "when." This Chapter presents a few of the many possible technical paths leading to a working molecular assembler that were being discussed in 1998. The reader is strongly urged to consult the most current literature for the latest results briefly summarizes the conventional top-down approaches to nanotechnology, describes several possible direct bottom-up pathways to molecular manufacturing including biotechnology, supramolecular chemistry, and scanning probes. Gimzewski and Joachim note that "bottom-up approaches to nanofabrication may one day compete with conventional top-down approaches in providing nanotechnologies for the next millenium. Top-down refers to increasing miniaturization through extension of existing microfabrication schemes, [whereas] the bottom-up scenario is one of ever-increasing complexity at the molecular level while maintaining control on an atom-by-atom basis". Surveys molecular mechanical components and design work on molecular assemblers and molecular manufacturing systems from the perspective of 1998.

Top-Down Approaches to Nanotechnology

Nanotechnology was first proposed by Nobel physicist Richard Feynman in December 1959, in a talk in which he also issued a seemingly "impossible" challenge to build a working electric motor no larger than a 1/64th-inch (400-micron) cube, backed by a $1000 prize to spur interest in the new field. Just 11 months later, engineer William McLellan had constructed a 250-microgram 2000-rpm motor out of 13 separate parts and collected his reward. (McLellan's entire motor is only as big as the period at the end of this sentence.) In 1995, a $250,000 Feynman Grand Prize was made available, this time sponsored by the Foresight Institute, for any engineer who could build the first programmable nanometer-scale robotic arm. How long will it take for history to repeat?

In his famous 1959 talk, Feynman proposed the prototypical top-down strategy for building complex nanomachinery—essentially a completely teleoperated machine shop, including mills, lathes, drills, presses, cutters, and the like, plus master-slave grippers to allow the human operator to move parts and materials around the workshop.

To build a nanomachine using Feynman's scheme, the operator first directs a macroscale machine shop to fabricate an exact copy of itself, but four times smaller in size. After this work is done, and all machines are verified to be working properly and as expected, the reduced-scale machine shop would be used to build a copy of itself, another factor of four smaller but a factor of 16 tinier than the original machine shop. This process of fabricating progressively smaller machine shops proceeds until a machine shop capable of manipulations at the nanoscale is produced. The final result is a nanomachine shop capable of reconstructing itself, or of producing any other useful nanoscale output product stream that is physically possible to manufacture, using molecular feedstock.

As Feynman describes the process: "When I make my first set of slave hands at one-fourth scale, I am going to make ten sets. I make ten sets of hands, and I wire them to my original levers so they each do exactly the same thing at the same time in parallel. Now, when I am making my new devices one-quarter again as small, I let each one manufacture ten copies, so that I would have a hundred hands at the 1/16th size.... If I made a billion little lathes, each 1/4000th the scale of a regular lathe, there are plenty of materials and space available because in the billion little ones there is less than two percent of the materials in one big lathe, it doesn't cost anything for materials, you see. So I want to build a billion tiny factories, models of each other, which are manufacturing simultaneously, drilling holes, stamping parts, and so on."

"As we go down in size, there are a number of interesting problems that arise. All things do not simply scale down in proportion. There is the problem that materials stick together by the molecular (van der Waals) attractions. There will be several problems of this nature that we will have to be ready to design for.... [But] if we go down far enough, all of our devices can be mass produced so that they are absolutely perfect copies of one another. We cannot build two large machines so that the dimensions are exactly the same. But if your machine is only 100 atoms high, you only have to get it correct to one-half of one percent to make sure the other machine is exactly tile same size—namely, 100 atoms high!"

Progress is being made on Feynman's top-down approach in a relatively new engineering field known as Micro Electromechanical Systems or MEMS, originally an extension of chip-etch technology rather than micromanipulation. Conventional purely-electronic devices fabricated on silicon chips with ~0.2 micron features must be coupled to external systems to produce mechanical effects, but MEMS devices integrate mechanical components directly with electrical circuitry. Thus while microelectronic chips merely route electrons, micro-electromechanical systems give the electronics immediate access to control applications, enabling single microdevices to interact directly with the physical world.

Generic microsystems research has proven to be a rewarding activity for two decades, with each successively smaller motor or machine widely publicized. The field gathered momentum after the first working micromotors were demonstrated in the late 1980s by groups at Berkeley and MIT. By 1990, tiny electrostatic motors with 100-micron rotors displayed operating speeds of ~250 Hz and operating lifetimes of ~10^6 revolutions. A new Journal of *Micromechanics* and *Microengineering* appeared in 1991, followed by the Journal of *Microelectromechanical Systems* in 1992 and the first MEMS

Application Symposium in Tokyo in June 1996. By the mid-1990s, Sandia's Microelectronics Development Laboratory, one of the most prominent vendors of micromachines, could mass-produce planar micromechanical devices smaller than ~1 mm^2, incorporating motors with gears and cogs each no thicker than a human hair and turning at speeds of >4000 Hz. Wobble micromotors such as those fabricated in silicon by the University of Utah's micromachine lab demonstrated very little friction or abrasion, with measured operating lifetimes in excess of 300 million revolutions. (Diamond coatings could further reduce friction by an order of magnitude over silicon, and the wear rate of polycrystalline diamond is known to be ~10^4 times lower than for silicon.)

MEMS research has produced multi-micron-scale accelerometers, diverse microsensors (e.g., blood pressure microsensors attached to cardiac catheters), microscale cantilevers and jointed crank mechanisms, 5-micron barbs, flow microvalves and pressure microtransducers, micropistons and micropumps, microgear trains, microactuators and piezodriven micromotors, micromirrors and microshutter Fresnel lens microarrays, microgyroscopes, a 2-mm long combustion chamber suitable for turbine use (making possible a 1 cm gas-turbine generator that would deliver 50 W of power or 0.2 N of thrust), and multidevice microsystems that were available customized or off-the-shelf in mass quantities by 1998. Laser beams and ion beams were used to carve ~300-micron-diameter gears with 50-micron wide teeth in solid diamond. Microelectrodeposition and microcontact printing readily produced topologically complex 3-D microstructures with 1-10 micron feature sizes. Colloidal templating yielded 0.71-micron diameter hollow silica spheres with wall thicknesses from tens to hundreds of nanometers, 10-25 nm trenches, or 4-nm thick graphite sheet layers tiled on spherical surfaces ~0.2 microns in diameter. A CAD/ CAM desktop micromanufacturing system invented at Lincoln Laboratory in Lexington, MA, machined three-dimensional structures, including spheres, in silicon, using computer-controlled laser/chemical etching of cubic-micron voxels at the rate of ~20,000 voxels/sec.

By 1998, smaller (~100 nm) accelerometers were being developed by such companies as Analog Devices in the U.S. and Daimler-Benz in Germany, and electronic control systems such as microswitches and microrelays were available commercially from several companies including Integrated Micromachines of California. Micromachining was a well-developed engineering discipline. Westinghouse Science and

Technology Center used MEMS sensors to build a miniaturized mass spectrometer. Micro-opto-mechanical systems were under development. Chemists were proposing and building Integrated Chemical Synthesizers consisting of millimeter- to micron-sized chemical reactors, designed for specific applications, along with associated devices for moving reactant and product streams (pumps), mixing reactants (reaction chambers), analyzing streams, and separating products (separation chambers), constructed conventionally using silicon substrates and photolithographic techniques to form the desired structures but also utilizing the high selectivity of biological molecules such as enzymes and antibodies to fabricate sensors. MEMS had also led to the development of microgrippers that could manipulate individual 2.7-micron polystyrene spheres, dried red blood cells of similar size, and various protozoa. Market Intelligence Research Corp. of Mountain View, CA, estimated a total worldwide budget for micromachine research and development of $995 million in 1992 and over $3 billion by 1998. Systems Planning Corporation of Arlington, VA, projected that the microsystems industry could be worth more than $14 billion annually by 2001.

Some MEMS research followed surprising directions. For example, in the mid-1990s MEMS engineers at the University of Tokyo Department of Mechanical Engineering fabricated an 8-hinge, 8-plate wing microstructure capable of independent flight through the air. The structure was fabricated by traditional silicon micromachining techniques and was then released from the wafer surface, with the initially flat sections lifted into flight position using a micromanipulation system with glass probes several microns in diameter. Each wing was ~500 microns square; applying 300 VAC at 10 KHz caused resonant vibration and the flapping amplitude of the wings exceeded 30°. In another 4-wing configuration, the flapping resonance frequency was ~150 Hz and the device freely flew up and down a thin glass pole when placed in an alternating electromagnetic field. Another bi-wing design made of magnetized nickel coated silicon measuring 2000 × 10,000 microns successfully flew without power supply cables and guides in the manner of a butterfly or mosquito, when an alternating magnetic field strength of >400 oersted and 12 Hz was applied; a similar device with 320 × 800 micron wings also flew in a 300 Hz magnetic field. "If we release the silicon mosquito from the silicon chip, it flies off and we cannot find it again," complained researcher H. Miura. "It's very small, like dust." MEMS "gnat robots" have also been described.

Using electron-beam (e-beam) lithography, in 1997 researchers at the Cornell University Nanofabrication Facility built what they believed were the world's smallest mechanical devices, including a 4-micron-wide Fabry-Perot interferometer and, just for fun, the world's smallest guitar, carved out of crystalline silicon and no larger than a single human cell. The "nanoguitar" was 10 microns long and 2 microns wide, with six strings each ~50 nm (~200 atoms) wide that would probably resonate at >10-100 MHz if plucked. Presentation of the nanoguitar enthralled the public, surprising at least one Cornell researcher (in another lab) who remarked that "anyone with a sufficiently advanced e-beam lithography system could make it. I make 50 nanometer silicon objects routinely. Doesn't seem like a big deal to me, but evidently most people have no idea what we're capable of these days."

Direct-write e-beam lithography uses a tightly-focused beam of electrons, steered by electromagnetic deflectors, to trace out patterns on a wafer resist surface. The electrons chemically alter the resist, which is then etched away, leaving the desired pattern. By 1998, e-beam spot sizes had been focused to less than 1 nm, and 5-nm beam widths were routine, although forward scattering of the electrons imposed a limit of resist exposure resolution of ~10 nm for e-beams and ~5 nm for focused ion beams. E-beam writing times were very slow—for example, 80 hours were required to carve a single photolithography mask for a 1G DRAM, though much of this time was required for proximity calculations by the AI (artificial intelligence) software (a better approach is to use a neural net). By 1998, an electron interferometer containing atomically smooth mirrors spaced a few atomic layers apart had been fabricated, and in 1999, careful electron-energy-loss spectroscopy experiments revealed that silicon dioxide must be at least 4 atoms (~0.7 nm) thick to act as a conventional electrical insulator.

Atom lithography, employing a low-energy neutral atomic beam (usually Na or Cr) cooled and collimated by optical molasses, then concentrated at the nodes of the standing wave formed by the superposition of a laser beam and its reflection, was predicted also eventually to allow surface deposition of 5-10 nm features. The first directional atomic beam (of sodium) with a ~0.002 radian beamspread and ~femtogram pulses (~26 million atoms, or ~(100 nm), per pulse) was demonstrated in early 1999, with one commentator noting that "the longer dream...is atomic holography...[that] could combine beams of atoms to build a 3-D solid object" At the same time, another group achieved a continuous stream of rubidium atoms lasting up to $\sim$0.1 sec in duration, with a beam radius potentially as small as $\sim$1 nm. Current-carrying wires have been used to guide cold lithium atoms "just as optical fibers guide light" the first curved focusing atom mirror also was demonstrated in early 1999. By 1998, the speculative possibility of quantum-mechanical micromanipulation also had been described.

Other approaches offered comparable results. For instance, engineers at the University of Minnesota Nanostructure Laboratory used nanoimprint lithography (NIL) in 1996 to etch out patterns of lines, grooves and circles as small as 25 nanometers in a polymer, and by 1998 NIL had been demonstrated in nanoscale quantum-wire, quantum-dot, and ring transistors with feature sizes below 10 nm. Nanochannel array glasses containing 33-nm capillaries arranged in a 2-D hexagonal close packing configuration at a number density of $\sim$300/micron^2 had been available since 1992.

Thus by 1998, top-down MEMS techniques could in theory already make nanoscale (though not atomically precise) parts. Might these techniques also produce assemblies of such parts, creating complex machines? Very simple mobile robots of $\sim$1 cm volume were commonplace, so for a more challenging demonstration of MEMS' ability to manufacture complete working microrobots, in 1994 Japanese researchers at Nippondenso Co., Ltd. fabricated a 1/1000th-scale working electric car. As small as a grain of rice, the microcar was a 1/1000-scale replica of the Toyota Motor Corp's first automobile, the 1936 Model AA sedan. The tiny vehicle incorporated 24 parts, including tires, wheels, axles, headlights and taillights, bumpers, a spare tire, and hubcaps carrying the company name inscribed in microscopic letters, all manually assembled using a mechanical micromanipulator of the type generally used for cell handling in biological research. In part because of this handcrafting, each microcar cost more to build than a full-size modern luxury automobile.

The Nippondenso microcar was 4800 microns long, 1800 microns wide, and 1800 microns high, consisting of a chassis, a shell body, and a 5-part electromagnetic step motor measuring 700 microns in diameter with a $\sim$0.07-tesla magnet penetrated by an axle 150 microns thick and 1900 microns long. Power was supplied through thin (18 micron) copper wires, carrying 20 mA at 3 volts. The motor developed a peak torque of 1.3×10^6 N-m (mean 7×10^7 N-m) at a mean frequency of ~100 Hz (peak frequency ~700 Hz), propelling the car forward across a level surface at a top speed of 10 cm/sec. Some internal wear of the rotating parts was visible after ~2000 sec of continuous operation; the addition of ~0.1 microgram of lubricant to the wheel microbearings caused the mechanism to seize due to lubricant viscosity. The microcar body was a 30-micron thick 20-milligram shell, fabricated with features as small as ~2 microns using modeling and casting, N/C machine cutting, mold etching, submicron diamond-powder polishing, and nickel and gold plating processes. Measured average roughness of machined and final polished surfaces was 130 nm and 26 nm, respectively. The shell captured all features as small as 2 mm on the original full-size automobile body. Each tire was 690 microns in diameter and 170 microns wide. The license plate was 10 microns thick, 380 microns wide and 190 microns high. Nippondenso subsequently used similar manufacturing techniques to build a prototype of a capsule intended to crawl through tiny pipes in a power plant or chemical plant like an inchworm, hunting for cracks. In 1999, three Japanese electronics companies announced the creation of a 0.42-gram, 5-mm long "ant size" robot reportedly able to lift 0.8-gram loads and to move at ~2 mm/sec, as part of the government's ongoing Micro Machine Project. By 1998, micromanipulators with high placement precision had been demonstrated in various laboratories around the world. For example, the Spider-II micromanipulation robot employed bimorphic piezoactuators within a 260-micron cubic work volume and a three-axis gripper placement accuracy of 8 nm. A microassembly robot using stick-and-slip actuators had 5-nm resolution over a 200-nm scanning range with a maximum speed of 4 mm/sec and 0.4% repeatability at 10 KHz, delivering a maximum driving force of 155 mN and rigidity of 6.3 N/micron.

In terms of system size, micromachined MEMS devices (typically $\sim 10^{-13}$ m^3 in volume) lie exactly intermediate between the world ($\sim 10^{-4}$ m^3) and the nanoscale world ($\sim 10^{-22}$ m^3). The medical nanorobots described in this book generally will have dimensions, precisions, and component part sizes ~1000-fold smaller than the Nippondenso microcar, which itself was, in turn, ~1000-fold smaller than a macroscale automobile. It is difficult to see how MEMS techniques involving statistical materials deposition and removal could fabricate parts to better than ~10 nm feature sizes or tolerances, or could position parts during assembly to much better than ~10 nm spatial resolution, nor does it appear likely that MEMS techniques, on their own, can manufacture or position any structure to atomic precision. However, this does not rule out the possible use of MEMS techniques to fabricate crude useful parts for subsequent "finishing" to atomic precision by other techniques, or to facilitate the assembly of nanoscale components (of atomic precision or otherwise) that have be.

Bottom-up Pathways to Molecular Manufacturing

To manufacture machines of any kind generally requires two primary capabilities—fabrication of parts, and assembly of parts. As will be shown below, by 1998 at least a primitive parts fabrication and parts assembly capability had been demonstrated at the molecular level using three different enabling technologies: biotechnology, supramolecular chemistry, and scanning probes. This Section reviews the most preliminary steps and basic approaches to molecular manufacturing. Consideration of more complex nanoscale components and assemblies is deferred to. The primary focus here is on bridging technologies that may lie between what can already be done and what is desired to be done—specifically, the manufacture of mechanical nanorobots. The reader is again cautioned that in 1998 this field was experiencing an explosion of interest and pathbreaking research efforts, so the current literature should be consulted for the latest results.

Biotechnology

The first broad category of enabling technology for molecular manufacturing in the mechanical tradition is biotechnology. Molecular biologists study and modify systems of molecular machines, and genetic engineers reprogram these systems, sometimes to build novel molecular objects having complex functions. By 1998, biotechnologists could make almost any DNA sequence or polypeptide chain desired (though not extremely long aperiodic ones), assemble "devices" such as artificial chromosomes and viruses, and were actively pursuing gene therapy, all of which are examples of molecular engineering. While earlier writers had suggested that biomolecules could be used as mechanical components, Drexler evidently was the first to point out, in 1981, that complex devices resembling biomolecular motors, actuators, bearings, and structural components could be combined to build versatile molecular machine systems analogous to machine systems in the macroscopic world. "Development of the ability to design protein molecules," Drexler wrote, "will open a path to the fabrication of devices to complex atomic specifications, thus sidestepping obstacles facing conventional microtechnology. This path will involve construction of molecular machinery able to position reactive groups to atomic precision." Perhaps the best-known biological example of such molecular machinery is the ribosome, the only freely programmable nanoscale assembler already in existence.

In nature, ~8000 nm ribosomes act as general-purpose factories building diverse varieties of proteins by bonding amino acids together in precise sequences under instructions provided by a strand of messenger RNA (mRNA) copied from the host DNA, powered by the decomposition of adenosine triphosphate (ATP) to adenosine diphosphate (ADP). Each ribosome is a compact ribonucleoprotein particle consisting of two subunits, with each subunit consisting of several proteins associated with a long RNA molecule (rRNA). Biotechnology can employ the ribosomal machinery of bacteria to produce novel proteins, which proteins then might serve as components of larger molecular structures.

Of course, such biological mechanisms have been disparaged as "slipshod devices working in spite of haphazard design." Dan Heidel, a University of Washington biomimetics engineer, agrees that biological systems are sloppy but observes that they are designed to tolerate it, and offers high praise for the lowly ribosome:

"The average ribosome sits in an aqueous environment surrounded by thousands of cosolutes at concentrations just a hair's breadth from precipitating out of solution. The ribosome itself consists of three RNA molecules that spontaneously fold into floppy spaghetti noodle piles of chemically active groups surrounded by >50 proteins of similarly dubious structure. The other solutes are bombarding this motley assemblage at tens to hundreds of miles an hour and the solvating water molecules are a constantly shifting and unpredictable network of hydrogen bonds and polar ionic charges constantly impinging upon the ribosome. The ribosome must recognize one particular cognate transfer RNA (tRNA) for the codon in its A-site out of ~60 other nearly identical tRNAs. This particular tRNA must be discriminated from all its tRNA brethren along with whatever other solutes can fit into the A-site pocket using only three base pairs for discrimination. Infact, most codons use only two of the three base pairs for recognition so some tRNAs are recognized by as few as 4 hydrogen bonds. All of this requires sub-Angstrom precision in the active sites but since the ribosome is a massive complex of subunits held together by only ionic, hydrogen, hydrophobic and Van der Waals bonds, it's like getting submillimeter accuracy from points on a bunch of taped-together water balloons being rolled down a staircase."

"This ridiculous machine can assemble proteins at a 20 Hz subunit-incorporation frequency with a 10^3 error rate. Ribosomes are even more impressive when one considers that the number of amino acids in each ribosome's own protein subunits combined with the 10^3 protein assembly error rate ensures that each ribosome has several errors in it and that each cell almost certainly has no two ribosomes exactly alike. I dare any engineer on this planet to rationally design an assembler at any scale that is able to operate at a comparable level of performance under similarly adverse operating conditions."

Thus ribosomes can manufacture, with >99.9% fidelity to arbitrary specifications, linear strings of amino acids of great length which may then fold up to produce three-dimensional protein structures performing a broad range of highly differentiated functions, providing good analogs for working nanomachines capable of forming self-assembling systems. Much of the great diversity of protein activity flows from the selectivity with which they bind to specific receptor sites on other molecules. Some of these binding mechanisms might be copied to construct molecular robotic arms, while other devices could be made by adapting naturally-occurring proteins, or by synthesizing new, custom-built ones. Protein engineering might be used to create proteins with desired qualities as building blocks for constructing the first primitive assemblers.

The field of protein engineering has its own journal of the same name, and has already achieved remarkable results in the synthesis of novel structures. Drexler cites the creation of the first de novo structure by DeGrado, the development of synthetic branched protein-like structures that depart substantially from protein models, and the engineering of a branched, nonbiological protein with enzymatic activity. Since then, numerous alpha-helical peptides have been designed de novo, a 20-residue artificial peptide exhibiting designed beta-sheet secondary structure has been created, and various peptide analogs have also been synthesized. By 1998, advances in computational techniques allowed the design of precise side-chain packing in proteins with naturally occurring backbone structures, and the study of de novo backbone structures had begun. The catalytic task space was being laboriously investigated at the molecular level—for example, one study of comparative enzyme structures revealed that changing as few as four amino acids converted an oleate 12-desaturase to an oleate hydroxylase, and as few as six substitutions could convert an oleate hydroxylase into an oleate desaturase. To further expand the toolkit.

Mills and Fahy proposed exploiting the degeneracy of the genetic code (wherein amino acids are specified by up to 6 different codons) by assigning some codons to unnatural amino acids, allowing the creation of artificial proteins with unprecedented structural or catalytic properties. With 61 usable codons available, there is room for up to 41 novel amino acids in addition to the natural 20 amino acids; unnatural amino acids have been introduced into beta-lactamase and T4 lysozyme enzymes in a site-directed fashion, using artificial tRNAs and mRNAs in an in vitro translation system, and into an engineered peptide as a reporter group.

Additionally, artificial novel base pairs can be incorporated enzymatically into DNA and other schemes are possible. By incorporating one additional base pair, the number of possible codons is expanded from $4^3 = 64$ codons to $6^3 = 216$ codons, potentially allowing up to 193 novel amino acids to be incorporated into artificial protein structures—or at least 48 new amino acids assuming that present levels of redundancy are retained and the existing 4-letter code is kept as a subset for "upward compatibility". If the genetic alphabet is increased to eight bases ($8^3 = 512$ codons), the number of available codons for novel amino acid-like molecules increases to as many as 489. Chains of amino acids shorter than ~100 residues are customarily called peptides; longer chains are proteins.

One of the greatest challenges in protein engineering (where one objective is to create functional, atomically-precise 3-D aperiodic structures) has been the difficulty of protein fold design, because individual amino acids have no strong, natural complementarity. In 1998, Duan and Kollman successfully folded a solvated 36-residue (~12,000-atom) protein fragment (i.e., a peptide) by molecular dynamics simulation into a structure that resembles an intermediate native state. During a 1-microsec simulation, the chain folded during 150 nanosec into a compact structure resembling the intermediate native state, as known by NMR, then unfolded and refolded again for a shorter period.

In 1998, whole protein folding to the final stable native state had not yet been computed deterministically using molecular dynamics—the real protein will fold and refold hundreds of times before it stumbles into the stable conformation with the lowest free energy—but research pursuing this objective was active and ongoing. Structural changes that occur during protein function (e.g., enzymatic action) were also being avidly studied. A different protein-based approach to near-atomically-precise 3-D structures avoids the folding problem by making use of semirigid protein nanoshells of deterministic size and shape to guide the ordered assembly of inorganic particles. For example, the capsid of the cowpea chlorotic mottle virus, a protein container with a 28-nm external diameter and an 18-nm internal cavity diameter, is composed of 180 identical coat protein subunits arranged on an icosahedral lattice. Each subunit presents at least nine basic residues (arginine and lysine) to the interior of the cavity, creating a positively charged interior surface which provides an interface for inorganic crystal nucleation and growth, while the outer capsid surface is not highly charged.

In one experiment, the empty capsid shell was found to act as a spatially selective nucleation catalyst in paratungstate mineralization, in addition to its role as a size- and shape-constrained reaction vessel. Noted the authors: "The range of viral morphology and size allow great flexibility for adapting this methodology to control the size and shape of the entrapped material, which is limited only by access to the protein interior and could include inorganic and organic species. The electrostatic environment within the cavity could be altered by site-directed mutagenesis to induce additional specific interactions."

It may also be possible to borrow various existing protein devices and apply them to new uses. The bacteriophage DNA injection system is self-assembling and could be engineered for other applications. Vogel has attached kinesin motors to flat surfaces in straight grooves; when ATP was added, the motors were activated, passing 25-nm wide microtubules hand over hand down the line in the manner of a ciliary array. Montemagno genetically engineered the 12-nm wide ATPase molecular rotary motor so that one end would adhere to a metal surface and the other end would provide an attachment site for 1-micron fluorescent streptavidin-coated bead payloads. When ATP was added, each bead individually attached to a biomotor in the motor array began twirling at ~10 Hz, generating a >100 pN force. Bead rotation was maintained continuously for more than 2 hours, indicating the high reliability of these ~100% efficient self-assembling biomotors. Montemagno explains that his laboratory's "long-term goal is the integration of the F_1 ATPase biological motor with nanoelectromechanical systems (NEMS) to create a new class of hybrid nanomechanical devices." Other natural nanomotors such as the self-assembling bacterial flagellar motor and the hexagonal "packaging RNA" or pRNA pumping mechanism that packs DNA into the capsid shell of the bacteriophage Phi 29 are also being intensively studied.

Yet another material which allows moderately well-defined three-dimensional nanoassemblies is DNA. The ideas behind DNA nanotechnology have been around since 1980, but activity in this field accelerated in the 1990s after numerous experimental difficulties were surmounted. Horn and Urdea reported branched and forked DNA polymers. Niemeyer and Smith exploited DNA specificity to generate regular protein arrays, and both suggested using self-assembled DNA as an early material for molecular nanotechnology. Shi and Bergstrom attached DNA single strands to rigid organic linkers, showing that cyclical forms of various sizes could be formed with these molecules. Mirkin's group attached DNA molecules to 13-nm colloidal góld particles with ~6 nm particle spacings, with the goal of assembling nanoparticles into macroscopic materials; Alivisatos, Schultz and colleagues used DNA to organize 1.4-nm gold nanocrystals in arrays with 2-10 nm spacings.

In either case, the specificity of DNA pairing should allow the construction of complex geometries since the use of colloidal particles can potentially add structural elements (which are more rigid than any polymer strand) to the set of building blocks for nanometer-scale structures—although to take full advantage of this capability would require particles which are atomically precise and which possess several chemically distinct anchoring points on each particle. Damha has synthesized V-shaped and Y-shaped branched RNA molecules, branching RNA dendrimers with "forked" and "lariat" shaped RNA intermediates, and even trihelical DNA. Henderson and coworkers have designed a simple DNA decamer that can form an extended linear staggered quadruplex array reaching lengths of >1000 nm, and have made branched oligonucleotides that template the synthesis of their branched "G-wires"; depending on the ratio of linear to branched building blocks, extensive DNA arrays with differing connectivities but irregular interstices can be created.

In 1998 the most intensive work on three-dimensional engineered DNA structures was taking place in Nadrian Seeman's laboratory in the New York University Department of Chemistry. Seeman originally conceived the idea of rigid 3-D DNA structures in the early 1980s, while examining DNA strands that had arranged themselves into unusual four-armed Holliday junctions. Seeman recognized that DNA had many advantages as a construction material for nanomechanical structures. First, each double-strand DNA with a single-strand overhang has a "sticky end," so the intermolecular interaction between two strands with sticky ends is readily programmed (due to base-pair specificity) and reliably predicted, and the local structure at the interface is known (sticky ends associate to form B-DNA). Second, arbitrary sequences are readily manufactured using conventional biotechnological techniques. Third, DNA can be manipulated and modified by a large variety of enzymes, including DNA ligase, restriction endonucleases, kinases and exonucleases. Fourth, DNA is a stiff polymer in 1-3 turn lengths and has an external code that can be read by proteins and nucleic acids.

During the 1980s, Seeman worked to develop strands of DNA that would zip themselves up into more and more complex shapes. Seeman made junctions with five and six arms, then squares, stick-figure cubes comprised of 480 nucleotides, and a truncated octahedron containing 2550 nucleotides and a molecular weight of ~790,000 daltons. The cubes were synthesized in solution, but Seeman switched to a solid-support-based methodology in 1992, greatly improving control by allowing construction of one edge at a time and isolating the growing objects from one another, allowing massively parallel construction of objects with far greater control of the synthesis sequence. By the mid-1990s, most Platonic (tetrahedron, cube, octahedron, dodecahedron, and icosahedron), Archimedean (e.g., truncated Platonics, semiregular prisms and prismoids, cuboctahedron, etc.), Catalan (linked rings and complex knots), and irregular polyhedra could be constructed as nanoscale DNA stick figures.

Seeman's DNA strands that formed the frame figures were strong enough to serve as girders in a molecular framework, but the junctions were too floppy. In 1993 Seeman discovered the more rigid antiparallel DNA "double crossover" motif, which in 1996 he used to design and build a stiff double junction to keep his structures from sagging. The next goal was to bring together a large number of stick figures to form large arrays or cage-shaped DNA crystals that could then be used as frameworks for the assembly of other molecules into pro-established patterns. These DNA molecules would serve as the scaffolding upon which new materials having precise molecular structure could be assembled.

In 1998, Erik Winfree and colleagues in Seeman's laboratory reported the design and construction of two-dimensional DNA crystals using self-assembling double crossover molecules. Repeating array units were approximately 2 nm × 4 nm × 16 nm in size, and examination of the array with an Atomic Force Microscope (AFM) revealed domains of up to 500,000 interconnected units, showing that the self-assembly process could be very reliable under ideal conditions. While the work described the construction of two- and four-unit lattices, the number of component tiles that could be used in the repeat unit did not appear to be limited to such small numbers, suggesting that complex patterns could be assembled into periodic arrays, which might also serve as templates for nanomechanical assembly. Noted the authors: "Because oligonucleotide synthesis can readily incorporate modified bases at arbitrary positions, it should be possible to control the structure within the periodic group by decoration with chemical groups, catalysts, enzymes and other proteins, metallic nanoclusters, conducting silver clusters, DNA enzymes, or other DNA nanostructures such as polyhedra." In these experiments, DNA hairpin turns incorporated within selected units inside their structures were visible as topographic features under AFM imaging, proving the ability to build predictable, atomically precise 2-D crystals with design control over every lattice point. Since the density of lattice points was much sparser than atomic spacings, this DNA self-assembly technology could be paired with atomically precise synthesis of diverse nanoscale subassemblies so that each unique double crossover unit in a pattern could be decorated with a unique subassembly of comparable size, possibly allowing the construction of mechanical nanocomputers and other nanomachinery.

In early 1999, Seeman reported yet another breakthrough—the construction of a mechanical DNA-based device that is a possible prototype for a nanoscale robotic arm. The mechanism has two rigid arms a few nanometers long that can be rotated between fixed positions by introducing a positively charged cobalt compound into the solution surrounding the molecules, causing the bridge region to be converted from the normal B-DNA structure to the unusual Z-DNA structure. The free ends of the arms shift position by ~2.0 nm during the structural conversion. Explained Seeman: "Using synthetic DNA as a building material, we have constructed a controllable molecular mechanical system. In the long-term, the work will have implications for the development of nanoscale robots and for molecular manufacturing."

DNA can also serve as an assembly jig in solution phase. Bruce Smith and colleagues are devising a method for the assembly and covalent linkage of proteins into specific orientations and arrangements as determined by the hybridization of DNA attached to the proteins, called DNA- Guided Assembly of Proteins (DGAP). In this method, multiple DNA sequences would be attached to specific positions on the surface of each protein, and complementary sequences would bind, forcing protein building blocks (possibly including biomolecular motors, structural protein fibers, antibodies, enzymes, or other existing functional proteins) together in specific desired combinations and configurations, which would then be stabilized by covalent interprotein linkages.

This technique could also be applied to nonprotein components that can be functionalized at multiple sites with site-specific DNA sequences, although proteins, at least initially, may be more convenient building blocks due to their size, their surface chemistry, the wide variety of functions and mechanical properties they can confer on the resulting assemblies, and the many existing techniques for introducing designed or artificially evolved modifications into natural proteins of known structure. Methods for covalently attaching functional proteins to a DNA backbone in a specified manner at ~8.5 nm (25 base-pair) intervals, addressable protein targeting in macromolecular assembly, and "protein stitchery" were being explored by others. Drexler notes that evolution has not maximized the stability of natural proteins, and that substantially greater stability may be engineered by various means (e.g., increasing folding stability by >100 Kcal/mole.

Molecular and Supramolecular Chemistry

A second broad category of enabling technology for molecular manufacturing includes molecular chemistry—the conscious design of completely artificial, nonbiological chemical structures that could potentially serve as molecular "parts" or which have specific device-like functionality. For example, catalysts may very loosely "be thought of as rudimentary assemblers that are slightly programmable through changes in the reaction milieu (i.e., changes in pH, temperature, etc.)". The number of natural molecular "parts" is immense—perhaps a few hundred thousand bioorganic compounds have been separated, purified and identified—but by 15 February 1999 chemists had already registered 19,245,458 well-characterized artificial molecules or other "substances" in the CAS Registry. Synthesis of natural molecules containing ~100 atoms (e.g., a ~1 nm molecular "part"), using methods of classical organic synthesis often requiring ~1 synthesis step per atom with ~90% yield per step, was state-of-the-art in 1998.

"What is exciting about modern nanotechnology," says Nobel chemist Roald Hoffmann, "is (a) the marriage of chemical synthetic talent with a direction provided by 'device-driven' ingenuity coming from engineering, and (b) a certain kind of courage provided by those incentives, to make arrays of atoms and molecules that ordinary, no, extraordinary chemists just wouldn't have thought of trying. Now they're pushed to do so. And of course they will. They can do anything." Molecular chemists study how small numbers of atoms combine covalently with each other to produce molecules with a wide variety of different properties.

Supramolecular chemists are primarily concerned with the much weaker noncovalent forces that can arise between molecules—such as hydrogen bonds and van der Waals interactions—that can be sufficient to bind collections of molecules tightly enough to form functional nanometer-sized structures. Supramolecular chemistry studies "chemistry beyond the molecule" or "the chemistry of the noncovalent bond". Its subject matter includes atomically precise supermolecules (well-defined discrete oligomolecular species resulting from the intermolecular association of a few components, such as a receptor and its substrate) and polymer-like supramolecular assemblies (polymolecular entities that result from the spontaneous association of a large undefined number of components).

Self-assembly is a key concept in supramolecular chemistry' as it is in nature, allowing the manufacture of large numbers of compound objects simultaneously and in parallel, rather than sequentially. Molecular self-assembly is a strategy for nanofabrication that involves designing molecules and supramolecular entities so that complementarity causes them to aggregate into desired structures. Self-assembly of atomically precise supermolecules thus demands well-defined adhesion between selected molecules, which may require steric (shape and size) complementarity, interactional complementarity, large contact areas, multiple interaction sites, and strong overall binding.

G.M. Whitesides explains that self-assembly has a number of advantages as a strategy. First, it carries out many of the most difficult steps in nanofabrication—those involving the smallest atomic-level modifications of structure—using the very highly developed techniques of synthetic chemistry. Second, it draws from the enormous wealth of examples in biology for inspiration; self-assembly is one of the most important strategies used in biology for the development of complex, functional structures, such as the well-studied examples of complex biomachine self-assembly of whole bacteriophage virions and flagellar rotor motors from their smaller molecular "parts". Third, self-assembly can incorporate biological structures directly as components in the final systems. Fourth, self-assembly requires target structures to be among the most thermodynamically stable available to the system, thus tends to produce structures that are relatively defect-free and self-healing. As Lehn points out: "By increasing the size of its entities, nanochemistry works its way upward towards microlithography and microphysical engineering, which, by further and further miniaturization, strive to produce ever smaller elements". Hierarchical self-assembly of many billions of micron-scale spherical components into a periodic honeycomb-structured photonic crystal optical device ~30 microns thick and ~1 cm^2 in area was demonstrated in 1998.

There is a wide range of different molecular systems that can self-assemble, including those which form ordered monomolecular structures by the coordination of molecules to surfaces, called self-assembled monolayers (SAMs), self-assembling thin films and Langmuir-Blodgett films, and self-organizing nanostructures. In these systems, a single layer of molecules affixed to a surface allows both thickness and composition in the vertical axis to be adjusted to 0.1 nm by controlling the structure of the molecules comprising the monolayer, although control of in-plane dimensions to <100 nm is very difficult. Fluidic self-assembly of microscale parts and the dynamics of Brownian self-assembly have also been described, and the theory of designable self-assembling molecular machine structures is beginning to be addressed.

Another approach using solid-phase peptide synthesis and a massively convergent self-assembly process was employed by M.R. Ghadiri and colleagues, who designed and synthesized a number of self-assembling peptide-based nanotubes using cyclic peptides with an even number of alternating D- and L-amino acids for the building blocks of the nanotubes. The alternating stereochemistry of the cyclic peptides allowed all the side chains of the amino acids to be pointing outwards which would not be possible in an ordinary all L-cyclic peptide. In this conformation, the amide backbone can H-bond in a direction perpendicular to the plane of the cyclic peptide. The stacking of two cyclic peptides forms an H-bonding network resembling an antiparallel beta-sheet as is commonly found in natural proteins. The H-bonding lattice quickly propagates perpendicular to the plane of the cyclic peptide, forming a tubular microcrystalline structure with 0.75-nm pores, or, in another experiment, 1.3-nm pores. Another group of nanotubes was designed with a highly hydrophobic outer surface and a hydrophilic inner pore. These nanotubes were easily inserted into a lipid bilayer and have been shown to be highly efficient ion channels; nanotubes with slightly larger pores transport small molecules such as glucose as well. Stable flat nanodisks with diameters continuously (chemically) adjustable from 30-3000 nm have been self-assembled in surfactant solutions.

Self-assembled crystalline solids such as zeolites incompletely fill space, leaving substantial voids that may be occupied by solvent molecules or other guest molecules, producing solid-state host-guest complexes known as clathrates (from the Latin *clathratus*, meaning "enclosed by the bars of a grating"). MacNicol and coworkers reported the first rationally designed clathrate host in 1978, but the first true *de novo* design of an organic clàthrate was in 1991 and involved the formation of an extensive three-dimensional diamondlike porous lattice built from a single Tinkertoy-type subunit containing four tetrahedrally arrayed pyridone groups that acted as connectors to assemble the units together in a well-defined geometry. However, the crystal was held together only by weak hydrogen bonds and the links had a rotational degree of freedom, rendering the exact crystal molecular arrangement unpredictable. Engineered nanoporous molecular crystals can provide molecular-scale voids with controlled sizes, shapes, and embedded chemical environments at resolutions of 0.3-4.0 nm. Other self-assembling crystal structures have been described including the crystal engineering of diamondoid networks, template-directed colloidal crystallization or colloidal epitaxy, organic templating to form crystalline zeolite-type structures with ordering lengths <3 nm, 3-D polymer channels with chemically functionalizable channel linings, and micromolding combined with templating and cooperative assembly of block copolymers to produce hierarchical ordering over discrete and tunable characteristic length scales of ~ 10 nm, ~ 100 nm, and ~ 1000 nm in a single body.

Dendrimers, also known as arborols or fractal polymers, are a well-known set of self-assembling structures that possibly could be used as tools to assist in the assembly of early nanomachines. Dendrimers are large regularly-branching macromolecules resembling fractal patterns, made by an iterative process in which small linear molecules are allowed to bind to each other at a certain number of sites along their length, building up branches upon branches with each iteration working outward from a core molecule having at least two chemically reactive arms, somewhat mimicking tree growth. Different core molecules or building-block chains produce macromolecules with different shapes, and the outer surface can be terminated with chemical groups of specific functionality, producing hundreds of differently-surfaced branched molecules that have already found use in medical research.

Growth is regulated, so size can be accurately controlled—dendrimers are typically a few nanometers wide but have been constructed with masses exceeding a million protons ($\sim 10^5$ atoms) and with diameters larger than 30 nm. In 1999, convergent assembly of specified 6-mer dendrimers, via trimerization of precursor dimer "parts," was demonstrated.

Autocatalysis or self-replication and is yet another approach to the chemical self-assembly of subunits. In 1989, J. Rebek reported the synthesis of super-molecules that could generate copies of themselves when placed in a sea of simpler molecular parts, with each component part consisting of up to a few dozen atoms. Molecular self-replication has continued to be studied by others. Ghadiri reported a self-replicating peptide in 1996, and by 1998 his laboratory had moved on to studies of more complex autocatalytic cycle networks. It is unknown how complex such assembly cycles may be made, but Ghadiri was investigating a self-replicating 256-component molecular ecosystem incorporating ~32,000 possible binary interactions, in 1998.

By 1998, chemists had already synthesized a vast variety of simple molecular parts of which only a tiny sampling can be mentioned here. For instance, carbon rings have been linked edge-on or by corners to create propeller-shaped molecules called *propellanes* or *rotanes*, or stacked with short hydrocarbon links to make gear-looking molecules such as cyclophane and superphane. Hydrocarbon molecular polyhedra in the shape of triangular (e.g., triangulane), cubic (e.g., cubane, first made by P. Eaton in 1964), pentagonal and hexagonal prisms, collectively known as *prismanes*, have been created, along with more unusual geometrical forms such as paddlane, housane, basketane, churchane, pagodane, and bivalvane; one carbon atom at each vertex). Cryptands, spherands, cryptaspherands and carcerands, corannulene baskets, and calixarenes are bowl-shaped molecules whose benzene ring walls hold their cavities rigid. The bowl rim of a calixarene can be lined with chemical groups to determine which guests it will accept, or two bowls can be bonded together, making a hollow nanoscale reaction vessel, and various molecular cages and capsules with precise sizes and characteristics have been assembled. A variety of nonchiral and chiral square molecules have been synthesized, as well as molecular "bottlebrushes"; DNA-polymer constant-force springs; the ferric wheel; molecular barrels and molecular tennis balls; various helical, boxlike, and other hollow organic structures; molecular "turnstiles" and "molecular tweezers"; simple and complex trefoil knot molecules; and stable carbyne molecular rods with chain lengths up to ~300 carbon atoms having alternating single and triple covalent bonds. Chemists are justifiably awed by the creation of structures like the ferric wheel; "for me," observed Nobel chemist Roald Hoffmann, "this molecule provides a spiritual high. ...The molecule is beautiful because its symmetry reaches directly into the soul, [playing] a note on a Platonic ideal." In chemical nomenclature, hydrocarbons with single carbon-carbon bonds are "-anes," double bonds make "- enes," and triple bonds make "-ynes.

Molecular gear systems first became a popular object of chemical study during the 1980s. For example, G. Yamamoto described "compounds that exist in conformations which are regarded as static meshed gears with two-toothed and three-toothed wheels and some of them behave as dynamic gears." H. Iwamura prepared one system that formed a chain of beveled molecular gears with ~GHz rotation rates, and another "doubly connected bevel gear system...[Transfer] of information from one end of the molecular system to the other end could take place in large molecules via cooperativity of the torsional motions of the chain." Kurt Mislow gave numerous examples of molecular gear systems that he had synthesized using the methods of traditional solution chemistry, which "resemble to an astonishing degree the coupled rotations of macroscopic mechanical gears." He added: "It is possible to imagine a role for these and similar mechanical devices, molecules with tiny gears, motors, levers, etc., in the nanotechnology of the future."

In 1998, Girnzewski and coworkers synthesized and observed a 1.5-nm diameter gear-like single molecule (hexatertbutyl decacyclene) rotating within a supramolecular bearing (a void in a 2-D lattice) at room temperature on the surface of a Cu{100} lattice. The authors suggest that added nonthermal noise (e.g., temperature differences produced by tunnel current heating of the rotor) could be rectified by asymmetries in the rotor/neighbour potential energy curve, allowing the rotor to turn unidirectionally. Says Gimzewski: "Our wheel is frictionless in the conventional sense of the word, and it doesn't wear out." Girnzewski's experimental work confirms Drexler's calculations that van der Waals bearings can operate in molecular systems, that multiatomic bearing surfaces with higher load capacities than sigma bond bearings can be built, that these bearings can run with no lubricants, and that such bearings have sufficiently low energy barriers to allow turning by thermal vibrations alone.

By 1998, limited control of the mechanical action of molecular parts had been demonstrated. For example, in 1994 T. Ross Kelly's group created a "paddlewheel" molecule (a spinning propeller-shaped wheel) with a built-in chemically-controlled brake. In solution, the wheel spins freely, but when Hg^{++} ions are added, a blocking ligand physically rotates into a new position (tripping the brake) and stopping the rotation; removal of the mercury causes the wheel to spinning. In 1997, Kelly's group used similar techniques to synthesize the first molecular ratchet, a propeller with three benzene ring "blades" that serve as gear teeth. A row of four rings (the pawl) sits between two of the blades, such that the propeller cannot turn without pushing it aside; because of a twist in the pawl, it is easier to turn clockwise than counterclockwise under thermal agitation, providing mechanical rectification without net motion or energy extraction. Shinkai and colleagues fabricated a pair of molecular tongs for grasping metal ions that uses two crown ether rings as the jaws, linked by two benzene rings joined through two double-bonded nitrogen atoms. Irradiation with ultraviolet light induces photoisomerization of the double bond between the nitrogen atoms, closing the jaws; heat reverses the process, opening the jaws. In related systems, collectively called "butterfly molecules," irradiation with visible light or pH changes can also induce opening. Seeman's chemically-operated mechanical DNA system provides similar control of molecular mechanical movement. Numerous controllable molecular electronics devices and possible nanocomputer components are reviewed in.

Noncovalent self-assembly of molecular parts has also been demonstrated. For example, Fraser Stoddart and colleagues have extensively investigated the rotaxanes, molecules in which one part is threaded through a hole or loop in another, and kept from unthreading by end-groups, like a ring trapped on a barbell. In one system, a ring-shaped molecule slides freely along a shaft-like chain molecule, moving back and forth between stations at either end with a frequency of ~500 Hz, making an oscillating "molecular shuttle". This shuttling behaviour can be controlled by a series of different chemical, electrochemical, or photochemical external stimuli. Using a different approach, G. Wenz threaded ~120 molecular beads onto a single, long, poly-iminooligomethylene polymer chain, making a "molecular necklace." Other groups are pursuing related research regarding [n]rotaxanes and pseudorotaxanes (mechanically-threaded molecules), while still others are using the necklace technique as a means, for example, of self-assembling ~2 nm wide cyclodextrin nanotubes.

Another interesting molecular system that can be made entirely by self-assembly is the catenanes, which have two or more closed rings joined like the links of a chain. The rings are mechanically linked—there are no covalent bonds between separate links. The first [2]catenane (2 linked rings) was constructed by Edel Wasserman of AT&T Bell Laboratories in 1960, using simple hydrocarbon rings; the first [3]catenane was synthesized in 1977. In 1994, Fraser Stoddart and David Amabilino constructed the first [5]catenane, having ~372 atoms, dubbed "olympiadane" because of its resemblance to the Olympic rings, and in 1997 the same group announced the first deterministically-ordered 7-ring heptacatenane; another group later reported the spontaneous self-assembly of a 10-component catenane. By 1998, much research on larger polymeric chains of linked rings (e.g., oligocatenanes), supramolecular "daisy chains", fullerene-containing catenanes, and supramolecular weaving was in progress. The smallest [2]catenane constructed to date has dimensions 0.4 nm × 0.6 nm. Self-assembled mechanically-interlocked 2-dimensional and 3-dimensional "infinite" arrays of single molecular ring-like species are known. Interestingly, in 1998 it was discovered that many viral coats (which also self-assemble) appear to be a 2-dimensional "chain-mail" weave of mechanically-interlocked protein rings—for instance, the spherical bacteriophage HK97 capsid shell consists of exactly 72 interlinked protein rings, specifically 60 hexamers and 12 pentamers.

Complex molecular parts also can be built up from simpler molecular parts by covalent bonding as low-symmetry shape-invariant molecular object polymers, or via processes variously known as *modular* chemistry, heterosupramolecular chemistry, chemical assembly, or structure-directed synthesis—including the well-known Diels-Alder reactions which may be employed repetitively to fabricate a series of atomically-precise molecular rods, rings, and nanocages (e.g., "beltenes" and "collarenes") composed of polyacenequinone units with nanometer-scale dimensions. Such efforts have been described as steps toward "molecular LEGO", "molecular Meccano", "molecular Tinkertoys", or "molecular building blocks". Some structure-directed assembly has been achieved with simple DNA forms. Other simple techniques for synthesizing nanowires, nanorods, nanotubes, and nanocages, or for selecting polymers for length, are well-known.

For instance, Jean-Marie Lehn has created rectangular grid complexes by mixing rodlike ligands that make use of different numbers of binding sites—one example is a 4 × 5 array consisting of nine ligands and 20 metal ions, nanoscale grids which he claims might one day find use "as components within a futuristic information storage and processing nanotechnology". C.M. Drain has described a 21-component (5 nm) square array of nine porphyrins tethered together by 12 palladium ions shaped like a four-pane window. The structure is built from three different kinds of porphyrins—(1) an X-shaped unit that coordinates to four metal ions, forming the center of the array, (2) a T-shaped unit that coordinates to three metals and forms each side of the array, and (3) an L-shaped unit that coordinates to only two metals and forms each corner of the array. When these components are placed in solution in the correct ratio, they form the square array within a half hour at room temperature with about 90% yield. The same porphyrin units can be combined in different ratios and induced to form wires or tapes. The properties of such arrays can be fine-tuned by choosing the appropriate metal ion linker and functionalized porphyrin unit.

Jeffrey Moore has investigated a three-dimensional nanoscaffolding comprised of a molecular lattice that could serve as a framework for catalysts, photosynthetic molecules, or more complex molecular devices. Moore's original objective was to synthesize modular building blocks with physical and chemical properties that would dictate a "programmed assembly" protocol for "nanoarchitectures". The characteristics of each modular unit shaped the weak intermolecular attractions—electrostatic, van der Waals, and hydrogen-bonding forces—between it and its neighbours so that units would array themselves in a larger structure with the desired spacing and geometry. Moore's group first linked phenylacetylene subunits into oligomers of various lengths which bend and twist like wire sculptures until their ends meet to form closed polyhedral molecules—the basic scaffold components (of which there were at least half a dozen shapes). These building blocks then self-assembled or folded into geometries at least partly controlled by designed hydrogen-bonding interactions. The molecular polyhedra each possessed several phenylene groups, "chemical handles" upon which other molecular groups could be hung. Moore then focused on ~10-nm dendrimeric forms and later began attaching photosensitive antenna molecules to these structures. By 1997, Moore had begun investigating norbornadiene on Si{100} in an effort to develop ways to covalently attach molecules onto surfaces with subnanometer spatial precision, and had suggested a packing model for interpenetrated diamondoid structures. Robson's group has also investigated possible scaffolding molecules.

Josef Michl is attempting to build a molecular construction kit using molecular rods and connectors, pursuing an explicit vision of "molecular Tinkertoys" by working with simple molecular structures that form stiff, flexible rods. Michl has assembled rods from a mixture of carbon-boron molecules (e.g., 10-vertex or 12-vertex carboranes) and carbon-hydrogen molecules, providing fine control over the total rod length. The rods are built up from more primitive molecular parts, such as propellane (a strained form of C_5H_6) and cubane (a strained form of C_8H_8), to make "staffanes" which are a series of cage units in a linear series. (Strained molecules are constructed with bonds that are forced out of their normal angles, as for example 90° in the case of cubane, compared with carbon's normal orientation of 109.5°.) Michl has fabricated rods whose lengths vary from 0.5 nm to 2.5 nm, in precise 0.1-nm steps. There are other ways of making rodlike molecules, but Michl's are highly inert, do not absorb visible or UV light, are stable up to at least 200°C, and do not react with the oxygen in air even at high temperatures.

Related work is underway in Michl's laboratory on connectors to join the rods together, Metal atoms would be the simplest solution, offering the useful quality of strong joints that can be easily disassembled. Different metals give different binding geometries—square, octahedral, and so on. Many of the connectors are metal-containing species; some are symmetrically trisubstituted or hexasubstituted benzenes or tetrasubstituted cyclobutadiene complexes. Michl must build the right chemical groups onto the rod termini to make them adhere to the connectors as desired. In one example, tiny crosses were built by attaching carboxylate groups to one end of the staffanes, allowing them to bind to a connector made of two rhodium atoms. The groups at the other end of the rods were converted to an ester group, which doesn't bind to rhodium, and so the crosses self-assembled correctly. Michl has also fabricated "star connectors" in which the rods are built into the molecules as covalently bonded arms of a star. For instance, one 3-arm star connector uses three large carboranes coupled to a central benzene ring, allowing a pedestal to be attached vertically to the benzene ring via a ruthenium "sandwich" bond. Additional devices could be installed on this molecular scaffolding, such as optically active molecules or even molecular mechanical "windmills". Nanoscale square and hexagonal planar grids have been fabricated. Extensive ab initio quantum mechanical calculations accompany the experimental work and aid in the interpretation of the results. Michl's ultimate goal is "the production of thin layers of solids of completely controlled aperiodic structure consisting of an inert covalent scaffolding carrying selected active groups."

G. Leach and colleagues used computers to simulate various diamond and graphite nanostruts using molecular dynamics calculations, analyzing solid rectangular struts with varying aspect ratios, cross-sections, terminating atoms (for diamond), potential energy functions, and temperatures. The most dramatic differences were seen between struts with 100:1 aspect ratios and struts with lower (10:1 or 1:1) aspect ratios—the diamond strut with a 100:1 aspect ratio and a unit cell (~0.4 nm) cross-section begins to curl (with end-to-end distance decreasing 1.2 nm after 20 picosec) due to thermal vibration (both at 150 K and 300 K), while a unit strut with an aspect ratio of 10:1 or a 100:1 strut with a (1 nm) cross-section showed end-to-end length fluctuations of only ~0.1 nm. The authors concluded that support struts with a cross-section of at least (1 nm) would be sufficient to keep both end-to-end distance and transverse fluctuations below ~0.1 nm in struts with 10:1 aspect ratios. Other work by the same group claims that many nanomachine designs may actually perform better than molecular dynamics results might suggest, or may require fewer atoms for the same positional stability. The elastic and wear characteristics of diamond are being studied computationally by others.

No discussion of molecular "parts" could be complete without mention of the fullerenes, first discovered in 1985 and vigorously investigated ever since. Fullerenes are one of the three known *allotropic* forms of carbon. In graphite, the most common allotrope, carbon atoms are arranged in hexagonal rings and strongly bonded into parallel planar sheets, with much weaker bonding between the sheets, giving graphite its excellent lubricating properties. In diamond, the second allotropic form, carbon atoms are arranged in a symmetric, tetrahedral structure, giving immense strength. Exposed to air, both diamond surfaces and graphite edges are quickly coated or "passivated" with hydrogen or other atoms that tie up the dangling bonds. In fullerenes, the third allotropic form, carbon atoms form large and hollow cage-like structures, often roughly spherical or tubular, made up of closed curved one-atom-thick sheets of carbon atoms arranged in a number of five-, six-, and higher-membered rings. Interestingly, the fullerenes need no passivating hydrogens or other atoms to satisfy their chemical bonding requirements at the surface—in this sense, "fullerenes are the first and only stable forms of pure, finite carbon".

A new journal, *Fullerene Science and Technology*, appeared in 1993, and by the mid-1990s ~3000 papers on fullerenes and their properties had been published and 149 fullerene-related patents had been issued in the U.S. alone (through 1996). By 1998, an incredible variety of fullerenes of many shapes and sizes had been synthesized.

The original fullerene, C_{60}, consists of exactly 60 carbon atoms arranged in a soccerball structure; lone carbon atom at each vertex) with 20 hexagons and 12 pentagons (with each pentagon entirely surrounded by hexagons). In theory, small molecules such as H_2 or CO, and probably CH_4, would fit inside C_{60}. A slightly larger variant, C_{70}, is the same except that an extra belt of carbon atoms has been inserted around the equator. Smaller and larger variants have been observed; as cages get bigger, the corners (where the 12 pentagons needed for closure reside) get sharper. The topologically smallest possible fullerene is a dodecahedron consisting of 12 pentagons and 20 carbon atoms, but the ring isomer of C_{20} (a hoop of single carbons) is apparently energetically favoured over the bowl or spherical (fullerene) isomers; in 1998, C_{28} was the smallest fullerene that had been observed experimentally, and C_{36} could be made in significant quantities by arc-discharge. The approximate diameter (measured to atomic centers) of a spheroidal fullerene of formula C_n is $D_{ball} \sim 1.1$ (n/60) nanometers. In 1998, purified C_{60} cost \$27.50/gm (99.5% pure) or \$60/gm (99.9%), C_{70} \$250/gm (98%), and C_{84} \$3,750/gm (90%).

Insertion of many additional equatorial belts of carbon atoms (incorporated as hexagons) into a spheroidal fullerene results in a long cylinder with graphite-like ("graphene") walls and spherical endcaps, making a class of fullerenes known as the capsular fullerenes or single-walled carbon nanotubes (SWNT) that come in many sizes and chiral forms. The molecular form of carbon nanotubes has been likened to rolled chicken wire. SWNTs may be synthesized by vaporizing graphite spiked with 1% catalyst from the nickel, cobalt and iron group above 3000°C, then allowing the vapor to slowly condense. Tubes form because the metal atoms interact with dangling bonds at the end of a tube, favouring tube extension over hemispherical capping by newly arriving carbon vapor atoms. (For any given mass number, ball-shaped fullerenes are energetically favoured over tube-shaped fullerenes during high-temperature synthesis in the absence of catalyst.) SWNTs typically self-assemble in 1.1 nm wide tubes, although smaller single-walled nanotubes and larger single-walled and multi-walled nested nanotubes exist. SWNT carbon atoms are bonded in virtually flawless hexagonal arrays. Simulations show that isolated flaws migrate to the ends of the tube and are eliminated, a phenomenon known as "self-healing". Single-molecule nanotubes of $C_{1,000,000}$ or larger, with lengths >1 00,000 times longer than their widths (e.g., ~1 mm long), had been synthesized by 1998. While graphite is a very brittle material, the graphene walls of a carbon nanotube are quite resilient. Computer simulations and experiments confirm that nanotubes kink when bent, then snap back when released. Nanotube diameter is $D_{tube} = 0.078\ (n^2 + nm + m^2)$ nanometers, where (n, m) is the "rollup vector" defined by the number of steps required for a repeat pattern along two crystallographic cylinder wall axes. The cohesive energy per atom required to curve a flat graphene sheet into a cylinderis $E_{rollup} = (13\ zJ\text{-}nm^2/D_{tube}{}^2)$. (1 zeptojoule (zJ) = 10^{21} joules.) As an example, a (10,10) nanotube has $D_{tube} = 1.36$ nm and $E_{rollup} = 7$ zJ/atom. Young's modulus for individual SWNTs, assuming a hollow (open-ring) cross-section, has been estimated as high as $\sim 5.5 \times 10^{12}\ N/m^2$; elastic bending modulus (measured as 1×10^{12} to $0.1 \times 10^{12}\ N/m^2$) decreases sharply with increasing diameter (from 8-40 nm). In 1998, purified carbon single-walled nanotubes were available via the Internet for $1400/gm from Tubes@Rice, or for $200/ gm at lesser purity from CarboLex.

From spatial geometry, it is known that hexagonal tiling produces flat sections (like cylinder walls), and inserting pentagons into an hexagonal array produces positive curvature (like spheres), making endcaps. But 7-sided heptagons can also be inserted into hexagonal arrays to induce negative curvature, thus permitting concave surface deformations such as saddle-shaped fullerenes or possibly helical tubular structures. Indeed, note Colbert and Smalley, "the use of hexagonal, pentagonal, and heptagonal substructures are sufficient to produce caged carbon structures of any topology". Nature makes use of this same geometrical principle in the radiolarians, tiny protozoans with fullerene-like siliceous skeletons, and in the conical nucleoprotein core particle of the HIV-1 virus. R. Smalley observes that a graphene sheet has the highest tensile strength of any known 2-dimensional network, and the packing density of atoms in the sheet (e.g., atoms/m) is higher than any other network made of any atoms in the periodic table and higher than the packing density in any 2-dimensional slice through any 3-dimensional object—even diamond which has the highest known 3-D packing density. Hence the graphene sheet is effectively impermeable under normal chemical conditions.

Carbon nanotubes have also been observed with a variety of endcap shapes and with tubes that reduced or increased their diameter for some distance before terminating. Toroidal fullerenes can be produced experimentally; cone-like, spindle-like, and helical fullerene objects have also been examined. By 1998, the mechanical properties of fullerenes and carbon nanotubes were being extensively investigated and there were experimental demonstrations of:

1. a limited ability to cut fullerene nanotubes to specific desired lengths;
2. dissolving derivatized SWNTs in common organic solvents;
3. the synthesis of self-oriented nanotube arrays; and
4. the trapping of individual C_{60} molecules in a perforated Langmuir-Blodgett film "workpiece holder" (though an even simpler means for producing 2-nm hole arrays with 78 nm hole spacings was later reported).

Nanotubes were also being investigated as nanoscale sensor components.

Given an ability to synthesize a wide variety of fullerene shapes and sizes, the next task is to find ways to join them together in specific desired molecular architectures. Two hydrogens were readily added to fullerene carbon atoms, making specific isomers of $C_{60}H_2$ and $C_{70}H_2$, and in the 1990s fullerene chemistry began to be explored. A little more than a decade after the discovery of C_{60}, gram quantities of buckyballs became widely available for chemical experimentation. Progress in covalent fullerene chemistry exploded. By the late 1990s, fullerene surfaces could be regioselectively functionalized in many interesting ways, including

1. a fullerene dimer,
2. a fullerene-polyester polymer,
3. a fullerene dendrimer,
4. a fullerene-rotaxane and catenane,
5. a fullerene-nucleotide DNA cleaving agent,
6. a Diels-Alder fullerene adduct that is very stable against cycloreversion, and
7. 3-dimensional extended polymeric multifullerene forms such as acetylenic macrorings and DNA/fullerene hybrid materials.

Theory and experiment suggest that SWNTs may be joined at 30° angles to create complex structures, including helices and three-way nanotube junctions.

In 1998, the idea of using fullerenes as "Tinkertoys" for molecular construction was just starting to be considered. For example, given the synthesis of C_{60} exohedral monoadducts and multiple adducts described above, it: may be possible to fabricate simple gears by bonding rigid ligands onto the external surfaces of carbon nanotubes in the manner of gear teeth, although the chemical synthesis of nanotubes with precisely positioned teeth will not be easy. Given the success in 1996 of IBM scientists in positioning individual organic molecules (each having a total of 173 atoms and a 1.5-nm diameter) at room temperature by purely mechanical means, it might also be possible to align and maintain these molecular gear teeth in atomically precise meshed positions. If this could be done, would such devices actually work in the manner of macroscale gears?

J. Han and colleagues at NASA/Ames performed a detailed 2000-atom molecular dynamics simulation to investigate the properties of molecular gears fashioned from carbon nanotubes with teeth added via a benzyne reaction known to occur with C_{60}. Computationally, one gear is powered by forcing the atoms near the end of the nanotube to rotate and a second gear is allowed to rotate by keeping the atoms near the end of its nanotube constrained to a cylinder (i.e., the ends of the shaft were constrained to not elongate but were allowed to move within a plane transverse to the tube symmetry axis). The meshing aromatic gear teeth transfer angular momentum from the powered gear to the driven gear. Each gear is made of a 1.1-nm diameter (14, 0) nanotube with seven benzyne teeth.

The spacing between two nanotubes is 1.8 nm and the smallest distance between a tooth atom and a tube atom is ~0.4 nm. The results show that the gears can operate up to 70 GHz in vacuo at room temperature without overheating or slipping. As rotational speed rises above 150 GHz, the gears overheat and begin to slip with tooth tilting up to 20°, but no bond or tooth breaking occurs up to at least 3000 K and slipping gears can always be returned to proper operation by lowering the temperature or the rotation rate. A related nanotube gear system (58 sprockets, 290-464 atoms) simulated by Robertson and colleagues at the Naval Research Laboratory (NRL) showed similar overheating at 500 GHz but was stable when accelerated to only 20 GHz.

The NASA group also simulated several other types of nanotube-based gear systems. For instance, a rack and pinion system was designed using a gear made from a (14, 0) nanotube with teeth separated by two hexagon rings and a shaft made from a (9, 9) tube with teeth separated by three rings. Gear and shaft are 1.94 nm apart, with tooth face normal to the radial direction of nanotube (14, 0) for the gear, but in the axis direction of nanotube (9, 9) for the shaft. The gear could receive power, driving the shaft, or vice versa, and worked well for shaft translational velocities up to ~100 m/sec. (Most devices described in Drexler typically move at 1 m/sec.) Since shaft mass is almost twice gear mass, it takes more power for the gear to drive the shaft. In another simulation involving a large 1.4-nm gear coupled to a smaller 0.8-nm gear, the large gear drives the smaller gear smoothly, but if power is instead applied to the smaller gear at a sufficiently large acceleration, then the smaller gear does not drive the larger one but instead "bounces back and forth several times, like elastic collisions of a small ball between two boards."

Another research group at Oak Ridge National Laboratory (ORNL) used classical molecular dynamics to investigate the properties of molecular bearings consisting of an inner and an outer carbon nanotube. The graphite bearings ranged in size from inner shafts between 0.4-1.6 nm in diameter up to 12 nm long, and outer cylinders between 1.0-2.3 nm in diameter up to 4 nm long. The original simulations found excessive vibrational motion, but subsequent work using a more complete quantum approach found that under certain conditions the nanobearing is "frictionless" and undergoes super-rotation, a classical dynamical behaviour reminiscent of superfluidity. The regime of superrotary motion is somewhat restricted when the nanobearing is under load, which suggests that a very careful design is required to ensure optimum performance.

The ORNL group simulated a fullerene motor consisting of two concentric graphite cylinders (shaft and sleeve) with one positive and one negative electric charge attached to the shaft. Rotational motion of the shaft was induced by applying one, or sometimes two, oscillating laser fields. The shaft cycled between periods of undesirable rotational pendulum-like behaviour and good unidirectional motor-like behaviour. The NASA/Ames group simulated a pulsed-laser-powered carbon nanotube gear-motor system which rotated consistently in one direction, although that direction could be either clockwise or counterclockwise.

G. Leach and colleagues also simulated various carbon nanotubes treated as nanostruts using molecular dynamics calculations, as described earlier for diamondoid block struts. The conclusions were similar though less severe. Nanotubes with 100:1 aspect ratio and 0.8-nm diameter at 300 K experienced periodic oscillations in length of ~0.4 nm at ~122 GHz, although multiple modes clearly were being excited; nanotubes of lesser aspect varied by at most ~0.04 nm.

Scanning Probe Technology

The third general pathway leading to molecular manufacturing involves a technology known as scanning probe microscopes (SPMs). The first of the SPMs was the Scanning Tunneling Microscope (STM) developed in the late 1970s and early 1980s by Gerd Karl Binnig and Heinrich Rohrer at an IBM research lab in Zurich, Switzerland, earning these scientists, along with Ernst Ruska, the 1986 Nobel in Physics. The STM was initially used as an imaging device, capable of resolving individual atoms by recording the quantum tunneling current that occurs when an extremely sharp conductive probe tip (usually tungsten, nickel, gold, or PtIr) is brought to within about one atomic diameter of an atom, and then adjusting the position of the tip to maintain a constant current as the tip is scanned over a bumpy atomic surface. A height change as small as 0.1 nm can cause tunneling current to double. The tip is connected to an arm that is moved in three dimensions by stiff ceramic piezoelectric transducers that provide subnanometer positional control. If the tip is atomically sharp, then the tunneling current is effectively confined to a region within ~0.1 nm of the point on the surface directly beneath the tip, thus the record of tip adjustments generates an atomic-scale topographic map of the surface. STM tips can scan samples at ~KHz frequencies, although slower scans are used for very rough surfaces. In some modern STMs (e.g., the DI Nanoscope), the sample is moved while the tip is held stationary.

A major limitation, of the STM was that it only worked with conducting materials such as metals or semiconductors, but not with insulators or biological structures such as DNA. To remedy this situation, in 1986 Binnig, Quate and Gerber developed the Atomic Force Microscope (AFM) which is sensitive directly to the forces between the tip and the sample, rather than a tunneling current. An AFM can operate in at least three modes. In "attractive" or non-contact mode (NC-AFM, 0.01-1 N/m force constant), the tip is held some tens of nanometers above the sample surface where it experiences the attractive combination of van der Waals, electrostatic, or magnetostatic forces. In "repulsive" or contact mode (CAFM, 0.01-1 N/m force constant), the tip is pressed close enough to the surface for tip and sample electron clouds to overlap, generating a repulsive electrostatic force of ~10 nN, much like the stylus riding a groove in a record player. There is also intermittent-contact mode (IC-AFM, 0.01-1 N/m force constant), which is sometimes called "tapping" mode. In any of these modes, a topographic map of the surface is generated by recording the up-and-down motions of the cantilever arm as the tip is scanned. These motions may be measured either by the deflection of a light spot reflected from a mirrored surface on the cantilever or by tiny changes in voltage generated by piezoelectric transducers attached to the moving cantilever arm. Typical AFM cantilevers have lengths of 100-400 microns, widths of 20-50 microns, and thicknesses between 0.4 to several microns. AFM tips may be positioned with ~0.01 nm precision, compressive loads as small as 1-10 pN are routinely measured, and the tips may be operated even in liquids.

S. Vetter notes that STM technology has also improved, reaching resolutions of -0.001 nm in the z direction (vertical) and ~0.01 nm in the xy plane, well beyond atomic resolution. The STM remains the instrument with the best resolution. The conducting surface limitation has been overcome in some cases by coating the target with an extremely thin conducting layer, developing tips with multiple electrodes, and coating a conducting substrate with a sample so thin as to allow enough conduction even if the sample is characterized as an insulator in bulk.

By 1998, the growing family of SPMs included at least forty types of instruments and techniques that relied on interactions between a scanned surface and a nearby probe. Different instruments measured different forces and thus could be used to characterize different properties of the surface. For example, friction force microscopes (FFMs), magnetic force microscopes (MFMs), shear force microscopes (ShFMs), scanning capacitance microscopes (SCMs), scanning conducting ion microscopes, chemical force microscopes, and electrostatic force microscopes (EFMs) measured frictional drag or other binding forces. Magnetic resonance force microscopes (MRFM) used a field generated from a small magnet mounted on the tip of the cantilever arm to probe nuclear magnetic moments in a small region on the surface of the sample, imaging atom types and even detecting the spin of a single electron. By the mid-1990s, AFMs were already a $100 million/year industry and SPMs generally were an off-the-shelf technology costing up to $50,000-$500,000 for complete systems, with the whole industry worth up to ~$0.5 billion annually. (Low-performance "homebrew," student, and science-fair STMs have been built for as little as $50.)

How might SPMs be used for molecular manufacturing? In the crudest approach, SPMs might be employed as nanoscale milling machines to carve out "nanoparts" from appropriate substrates. For example, Kim and Lieber used an AFM to perform nanomachining operations on a molybdenum trioxide crystal—applying a 100 nN load at the tip, they milled a triangular-shaped 50-nm planar chunk from the crystal, then slid the part 200 nm across the work surface by pushing it with the AFM tip. Sheehan and Lieber milled a 50-nm rectangular part and two other parts with concavities complementary to the rectangle by mechanically etching a MoO_3 layer deposited atop an MoS_2 surface by pressing down with an AFM tip at a 50 nN load. Like a numerical-controlled machine tool, the AFM motions could be programmed to perform a series of steps, like automatically carving the chemical formula "MoO_3" into the crystal. Nanoparts with features as narrow as 10 nm can be nanomilled in a limited set of materials. In theory, large arrays of independently-controlled SPM tips could be operated in parallel to carve out great numbers of similar objects simultaneously, though such objects would of course not be atomically identical without further "finishing." AFM tips have also been used to bulldoze nanoscale lines and rectangular <10-nm features into nonconducting photoresist, perform "dip-pen" AFM nanolithography, form grooves, and fabricate a single-electron transistor via an STM nano-oxidation process. A carbon nanotube affixed to an AFM tip was used as a pencil to write 10-nm-wide features onto silicon substrates at ~0.5 mm/sec: An electric field issues from the nanotube, removing hydrogen atoms from a hydrogen monolayer atop a silicon base; the exposed silicon surface oxidized, producing narrow SiO_2 tracks.

Perhaps more significantly, SPMs have been employed in an increasingly sophisticated manner to manipulate individual atoms and small clusters of atoms. The possibility of modifying surfaces with scanning probe tips was evident from the earliest years of STM research, since inadvertent contact between tips and surfaces routinely caused such modifications, and the possibility of transferring material between tip and sample had already been discussed. Suggestions for controlled surface modification soon appeared, and in 1985 Becker and Golovchencko used voltage pulses on an STM tip to pluck a single germanium atom from the surface of a sample. In 1988, J. Foster and colleagues at IBM Almaden pinned small organic molecules to a graphite surface by applying a small electrical pulse through an STM. Additional pulses enlarged or erased this molecular feature, and often split the organic molecule into smaller pieces, though not in a controlled manner. Other workers used an STM to dump small clusters of gold atoms on a platinum surface, clusters of copper atoms on a gold surface, and molecules of tungsten carbide from a supply flowing over a surface.

Others reported positioning carbon monoxide molecules and platinum atoms on platinum surfaces, moving clusters and single atoms of silicon across a silicon surface at room temperature, and fabricating via STM a "molecular corral" or "quantum corral"—a ring of atoms so small that the enclosed electrons were forced to exhibit quantum behaviour. In 1999, Tomanek and Kral proposed an atomic fountain pen in which a carbon nanotube filled with atoms is electrically induced to release these atoms, one by one every 15 microsec, onto the work surface.

More precise control was achieved by Eigler and Schweizer at IBM Almaden in 1989, when they used an STM to position 35 individual xenon atoms on a nickel surface to spell out the corporate logo "IBM". To accomplish this, a bias voltage was applied to weaken the adsorption of each atom to its nickel substrate, then each atom was dragged, one by one, to the desired locations at a speed of ~0.4 nm/sec to build a meaningful pattern. It took 22 hours to make the entire logo, or ~38 min/atom. The experiment had to be carried out at 4 K (i.e., liquid helium temperatures) because the arrangement would have been unstable at room temperature.

Numerous similar examples of "atomic graffiti" followed the IBM experiment, including atoms patterned in the shape of a world map; the word "atom" and "nanoworld", spelled out with atoms, in Japanese; the word "Peace," Einstein's famous equation "$E = mc^2$," and even a reproduction of the well-known Einstein portrait with the scientist's tongue sticking out; a sketch of a "molecule man" using 28 individual CO molecules on a platinum surface; gold nanoparticles spelling out "USC" and "Zyvex"; and sulfur atoms removed from a surface to make letters 2 nm high. In 1996, Gimzewski's group used an STM tip to manipulate individual buckyballs along terraces on a grooved copper plate, making a "molecular abacus."

Some variant of an SPM would appear to be the most promising tool for direct molecular manipulation. Once a tip lies within ~1 nm of a surface, the potential barrier can be lowered sufficiently for atoms to be induced from the surface by field evaporation; an applied tip voltage ionizes the atom which can then be guided around by the tip. Depending on voltage and separation, atoms, atom clusters, or molecules can be pried out, pushed in, or nudged around on a surface.

By 1998, techniques for direct atomic manipulation proceeded at room temperature and ~1000 times faster than the original IBM logo had been assembled. For example, the Nanomanipulator, an interactive haptic control system created by a group at the University of North

Carolina at Chapel Hill, allows near-real-time manipulation of individual gold atoms across a surface using a hand-held master-slave controller that drives an STM probe while the position of the atom is displayed on a monitoring screen visible to the user. Partially funded by NSF, Silicon Graphics, and TopoMetrix, the Nanomanipulator is a fully-integrated system that enables investigators to "feel" the interatomic forces as the user pushes atoms around on a surface. In one demonstration, a TV-watching user horded a gold atom, slipping and sliding, into a slot in the planar workplace in about 1 minute; when force feedback was turned off, movements tended to run wild. IBM Almaden has also experimented with removing individual atoms from metal surfaces using an STM hooked into a Virtual Reality Dataglove apparatus.

To manufacture atomically precise parts, it also will be necessary to manipulate covalent bonds at the probe tip. STMs have broken and created chemical bonds; ~1 volt pulses were used to pull atoms out of crystals, binding them to the tip, and then to reinsert them back into the crystal. In 1995, the first demonstration of catalysis on a nanometer scale was reported by scientists at the Molecular Design Institute at LBL. They used an AFM modified to function like an ultrafine-point pen for catalytic calligraphy to change the chemical composition of a material surface one molecule at a time. A surface was prepared as a self-assembled monolayer (SAM) of alkylazide molecules capped with a crown of three nitrogen atoms, then platinum-coated chromium was deposited onto an AFM silicon tip just a few atoms wide. The SAM was soaked with a hydrogen-containing solvent, then scanned by the AFM over a 100 micron area, with the platinum catalyzing a covalent bonding reaction in which hydrogen was added to the azides, transforming them into amines as revealed by selective fluorescent tags.

Tip technology is an important and fast-growing subspecialty of SPM research. In 1990, Drexler and Foster suggested the use of custom-made, synthetic proteins to be mounted on the point of an SPM tip. Modified antibodies or specially designed proteins could serve as simple, first-generation "grippers" for binding and manipulating specific molecules, bringing them into position to react with other molecules in a precise and selective way.

Functionalized SPM tips have been created to exploit antibody-antigen recognition and in the context of chemical force microscopy. A major convenience for molecular manipulation is that the same instrument that does the chemistry and the molecular orienting can also be used to inspect the results. In 1996, Smalley's group at Rice University attached a single nanotube to the pyramidal silicon tip of an AFM and showed that it was quite robust and could image the bottom of deep trenches inaccessible to conventional tips. By 1998, progress had been made in functionalized carbon nanotubes for use on AFM tips.

Most notably, Wong and colleagues prepared nanotube tips by oxidation in air at 700°C, burning off all but 2% of the original material and leaving the ends covered with carboxyl (COOH) groups whose chemistry is rich and well understood. Four different kinds of tips were created: (1) the original carboxyl tip, which is acidic; (2) an amine-terminated tip (made by forming an amide bond to one of the amine groups in ethylenediamine (H_2NCH_2,CH_2NH_2), which is basic; (3) a hydrocarbon-terminated tip (made by forming an amide bond to benzylamine ($C_6H_5CH_2NH_2$), which is hydrophobic; and (4) a biotin-terminated tip (made by forming an amide bond to a biotin derivative, which shows specific binding to streptavidin. AFM contact forces between tips and selected substrates were shown to be sensitive to pH and to the chemical details of the substrate in ways consistent with the tips' intended chemistry.

Wong's tips had three closely related advantages over previous techniques. First, when a functional group is attached to the apex of an Si_3N_4 or SiO_2 tip, these groups usually adhere to the sides of the tip as well; upon using the tip as a tool, there is a constant hazard that contact with the sides of the tip will alter the workpiece in unwanted places. Unlike these tips built with bulk techniques, the ends of Wong's tips are very different from their sides—the carboxyl groups are attached only on the ends of the nanotube, not on the sides. Second, Wong's tips have lateral dimensions set by the nanoscale dimensions of nanotubes, not by top-down fabrication techniques. Note the authors: "...the small effective radius of nanotube tips significantly improves resolution beyond what can be achieved using commercial silicon tips....we have recently demonstrated that lateral resolution of <3 nm can be achieved by using COOH-terminated single-walled nanotubes tips". Third, single-walled nanotube tips are much closer to yielding truly single atom tips with controlled chemistry than any other alternative. J. Soreff notes that a nanotube with a 1.4-nm diameter has just 20 atoms at an open end. Even given a statistical distribution of tubes with varying numbers of carboxyl groups attached to their ends, one should be able to build a ligand which covers the whole end of the tube, yielding a method for ensuring that just one molecule of known structure and orientation was present at the end of the probe.

Theoretical studies have employed molecular dynamics simulations and other techniques to investigate the behaviour of mechanosynthetic tool tips. For example, Drexler proposed using an acetylene radical to perform selective abstractions of hydrogen from a diamond surface and this proposal was further studied in Musgrave's ab initio quantum chemistry analysis and Sinnott's molecular dynamics modeling of the reaction at room temperature on a diamond surface; other H-abstraction studies have been done. One possible structure for a hydrogen abstraction tool might be similar to an anthracene whose end had been modified, prior to use, the structure must be activated by removal of the terminal hydrogen. Brenner and colleagues describe further work modeling the parallel abstraction of several hydrogen atoms from the diamond surface. Such reactions are similar to those involved in the well-studied chemical vapor deposition of diamond, but adding only positional control and applied mechanical force; more detailed theoretical studies of diamond mechanosynthesis reactions have begun.

R. Merkle has proposed a simple mechanosynthetic system capable of fabricating a large class of useful stiff hydrocarbons. Merkle's system includes positionally-controlled hydrogen abstraction and deposition tools (to remove or add hydrogen atoms to a workpiece), carbene and dimer deposition tools (to add one or two carbons to a workpiece), a silicon radical tool (to remove a carbon), and four other specialized tools. Each proposed tool has a molecular handle structure that could be permanently affixed to a suitably functionalized AFM tip (or other appropriate mechanosynthetic nanomanipulator). As each operation is performed on the butadiyne (a.k.a diacetylene, C_4H_2) hydrocarbon feedstock in an inert (e.g., vacuum or noble gas) environment, one or more tools are brought to the workpiece at the necessary position and angle, and any necessary force is applied via the AFM. Interestingly, except for a small number of "vitamin parts" involving transition metals, Merkle's system can in theory synthesize all of its own mechanosynthetic tools, thus establishing net positive production of system components and illustrating a limited subset-level "quantitative parts closure" as first defined in 1982 by Freitas and Gilbreath in connection with self-replicating machine systems. Merkle acknowledges that several of the proposed reactions involve the simultaneous coordinated action of two, three, or even four positionally controlled tools, but suggests that two 6-degree-of-freedom manipulators plus some combination of appropriate jigs and fixtures should be sufficient in a more parsimonious system. However, steric hindrance of multiple tools at one site also was not explicitly addressed. Another potential problem to be overcome is loss of positional registration as the various functionalized tool heads are changed out at the AFM tip; such tips might be paired with a second positionally-invariant metrology tip on each tool head, allowing precise calibration of workspace location before proceeding back to the workpiece.

Experimentalists are warned that butadiyne, chemically distinct from the more common molecule butadiene (C_4H_6), can polymerize explosively under some conditions. Others have suggested fabricating nanoparts by building up structures by stacking layers of individual atoms, perhaps constrained by sacrificial joiners and scaffolding, and honed by differential etching operations.

Assuming that "nanoparts" can be fabricated to atomic precision, can SPMs also be used to assemble these nanoparts into working nanomachines? Some preliminary work has been accomplished. For instance, arrays of nanoscale holes have been synthesized that could serve as "workpiece holders." In one experiment, C_{60} buckyballs were scattered about on a prepared surface, then an AFM pushed the individual spheres into their own little holes, securely seating them pending further processing. The C_{60}-impregnated perforated film was so durable that the buckyballs remained trapped after several months of storage. Similarly, Jung and colleagues in Gimzewski's group synthesized 4-legged porphyrin-based molecular "nanoparts" that were specifically designed for easy positioning on a copper surface. Pushing with an STM at room temperature, the authors displaced the ~1.5-nm wide nanoparts in predefined directions and rotated the 4-legged molecules at will, arranging, for example, the parts into a precise hexagonal configuration on the copper surface. The STM has been used to induce, and to view, the rotation of an individual oxygen molecule trapped on a platinum surface, as it is intentionally and stably rotated into any one of three distinct orientations. The forces required to slide, or alternatively to roll, a carbon nanotube across a graphitic surface have been directly measured by AFM.

The first known instance of a crude but purely mechanical "nanopart" assembly operation was performed in 1995 by Paul E. Sheehan and Charles M. Lieber. Two nanoscale parts with rectangular slots and a third 50-nm rectangular sliding "latch" member were milled from a MoO_3 crystal using an AFM, then the rectangular "latch" member was slid repeatedly from one slot to the other using the same AFM, making a three-nanopart reversible mechanical latch. Noted the authors: "The lateral force needed to break the latch, 41 nN, was large considering the small latch contact area, which suggests that relatively robust assemblies can be created with such devices. Such a reversible latch could serve as the basis for mechanical logic gates....Our results..demonstrate the ability to machine complex shapes and to reversibly assemble these pieces into interlocking structures."

In 1998, manipulation of nanoscale parts in the vertical dimension (e.g., normal to the surface plane) had only just begun. The first three-dimensional structure built out of single "nanoparts" was demonstrated in 1997 by Requicha's group at the USC Laboratory for Molecular Robotics. First, an AFM was used to push 5-nm gold nanoparticles in the third dimension up and over surface protrusions or obstructions—for example, a 5-nm gold particle was shoved up onto a 2-nm-high protrusion, then slid back off again onto the surface. In another series of experiments conducted in air at room temperature with 15-30 nm gold nanoparticles deposited on silicon previously coated with a silane layer, Requicha's group used an AFM to build a simple 3-D pyramidal structure by pushing one gold "nanopart" on top of two others, then off again; the group also used the AFM to rotate and translate a dimer unit formed by two linked "nanoparts".

Three-dimensional nanoassembly might perhaps be more readily achieved if nanoparts could assist in their own assembly process. R. Merkle suggests that an SPM could be used to position DNA-tagged molecular building blocks. A building block tagged with a specific single-stranded DNA should preferentially adhere to a surface covered with the complementary DNA. If the building block has a second single-stranded DNA tag which is complementary to single-stranded DNA on the tip of an SPM, and if it is initially attached to the SPM tip by this tag, then it could be positioned by the SPM and attached to the surface. The relative strength of attachments could be adjusted by changing the number of base pairs in the complementary region between the two strands. A related suggestion, due to G. Fahy, is that SPM tips for assembly could be designed with more than one binding site. The existence of at least two binding sites could allow precise orientation of a nanopart, permitting the nanopart to be presented to a desired target site (or another nanopart, possibly attached to another manipulator) at exactly the location and in exactly the orientation desired, assuming that each nanopart can bind to each tip in only one possible orientation. A convenient post-assembly nanopart release mechanism is also required.

In 1997, the Avouris group at IBM Yorktown Heights demonstrated that the tip of an atomic force microscope (AFM) could be used to control the shape and position of individual multi-walled carbon nanotubes dispersed on a surface. Nanotubes could be bent, straightened, translated, rotated, and (under certain conditions) cut.

In 1998, Zyvex LLC (a developmental engineering company whose goal is to create a molecular assembler capable of manufacturing atomically precise structures) demonstrated the ability to manipulate carbon nanotubes in three dimensions inside a scanning electron microscope (SEM) having ~6 nm resolution at near-video scan rates. Zyvex's custom piezoelectric vacuum manipulator achieved positional resolutions comparable to SPMs along with the ability to manipulate objects along one rotational and three linear degrees of freedom with 0.1 nm spatial resolution. This prototypical device could probe, select and handle limited classes of nanometer-scale objects such as carbon nanotubes under real-time SEM inspection. Carbon nanotubes were attached to commercial atomic force microscope (AFM) tips either by van der Waals forces alone or by "nanosoldering" by a concentrated electron beam from the SEM. Forces applied to the nanotubes could be measured from the cantilevers' deflections (spring constants 0.01-100 N/m). Sometimes nanotubes were transferred from tip to tip. The manipulator could function both as a research tool for investigating properties of carbon nanotubes and other nanoscale objects without surface restrictions, and as a rudimentary building device for larger nanotube assemblies. Zyvex believed that this capability to select and manipulate nanoscale components and to examine directly their suitability as construction materials during various phases of the construction process would play an important role in enabling the technology of assembling mechanical and electronic devices from prefabricated components. Most impressively, the Zyvex system allowed the simultaneous coordinated operation of three independently controlled AFM tips within the same workspace, at different orientations. (A four-tip system was expected to be operational by mid-1999.) Two-handed coordinated micromanipulation under the view of an SEM had previously been demonstrated by others.

Tuzun and colleagues calculated the requirements for coaxial docking of two nanotubes of different diameters to form a molecular bearing. They looked at bearings formed from nanotubes II rings long, with 10 carbons per ring in the shaft and either 30 or 34 carbons per ring in the sleeve. For the computer simulations, Tuzun placed the sleeve of the bearing along the z axis and gave the shaft a small initial velocity towards it. A perfectly aligned shaft falls straight into a potential well from the van der Waals attraction to the sleeve; displaced or misoriented shafts can bounce off the edge of the sleeve. The docking envelopes for the molecular dynamics calculations and for the rigid body calculations had essentially the same shape, but with slightly different sizes.

For the atomistic calculation, a shaft aligned parallel to the sleeve (with a 30 carbon ring sleeve) can be displaced by 0.26 nm before it fails to dock, while in the rigid body calculation the displacement can only be 0.16 nm. The authors note that "if an end of the shaft points closely enough to the center of the sleeve, it will fall into the nonbonded potential energy well and the two nanotubes will dock." These calculations are important because they tell us what tolerances can be accepted during the manual assembly process. Note the authors: "A question just as important as how to design or operate nanomachines is how to assemble them". Interestingly, one of Zyvex's experiments produced a multi-walled carbon nanotube that thinned in three steps along its length, suggesting the most likely explanation that the inner shells had been pulled from the outer shells. This phenomenon has also been observed by others and is known as the "sword and sheath" failure—the exact inverse of the purposeful nanotube insertion operation investigated by Tuzun and colleagues.

In order to produce large numbers of nanoparts and nanoassemblies, massively parallel SPM arrays and microscale SPMs would be most convenient. Force sensing devices such as piezoelectric, piezoresistive, and capacitive micro-cantilevers made it possible to construct microscale AFMs on chips without an external deflection sensor. (We exclude fixed-tip arrays in the following discussion.) In 1995, Itoh and colleagues at the University of Tokyo fabricated an experimental piezoelectric ZnO_2 on SiO_2 microcantilever array of ten tips on a single silicon chip. Each cantilever tip lay ~70 microns from its neighbour, and measured 150 microns long, 50 microns wide and 3.5 microns thick, or ~26,000 micron/device, and each of the devices could be operated independently in the z-axis (e.g., vertically) up to near their mechanical resonance frequencies of 145-147 KHz at an actuation sensitivity of ~20 nm/volt—for instance, 0.3-nm resolution at 125 KHz.

Parallel probe scanning and lithography has been achieved by Quate's group at Stanford, which has progressed from simple piezoresistive microcantilever arrays with 5 tips spaced 100 microns apart and 0.04-nm resolution at 1 KHz but only one z-axis actuator for the whole array, to arrays with integrated sensors and actuators that allow parallel imaging and lithography with feedback and independent control of each of up to 16 tips, with scanning speeds up to 3 mm/sec using a piezoresistive sensor. By 1998, Quate's group had demonstrated arrays of 50-100 independently-controllable AFM probe tips mounted in 2-D patterns with 60 KHz resonances, including a 10 × 10 cantilevered tip array fabricated in closely spaced rows using throughwafer interconnects on a single chip.

MacDonald's group at the Cornell Nanofabrication Facility has pursued similar goals. In 1991, the team fabricated their first submicron stylus, driven in the xy plane using interdigitating MEMS comb drives, including the first opposable tip pair. By 1993, they had produced a 25-tip array on one xyz actuator, and by 1995 a complete working microSTM (including xy comb drives) measuring 200 microns on an edge and a microAFM measuring 2 mm on an edge including a 1-mm long cantilever with a 20-nm diameter integrated tip on a 6-micron high by 1-micron diameter support shaft.

MacDonald's group demonstrated tip arrays with 5 micron spacings, exploiting the same process used to make the working microSTM. With the same technology tips or small arrays of tips could be spaced 25-50 microns apart and integrated with individual z-axis microactuators, so that one xy-axis manipulator could support many tips with each tip having a separate z actuator.

By 1997, the group had built and tested an array of microSTMs on the surface of an ordinary silicon chip, with each tip on a cantilever 150 microns long with 3-D sensing and control. The largest prototype array has 144 probes, arranged in a square consisting of 12 rows of 12 probes each, with individual probe needles about 200 microns apart. Further development was to focus on increasing the range of movement and on fitting more and smaller probes into the same space.

Molecular Components and Molecular Assemblers

Assuming that various elementary molecular parts fabrication and assembly operations can be performed for certain limited classes of structures, perhaps as described in, then what structures would be interesting and useful to build? This Section briefly describes a small set of possibly useful molecular mechanical components and then concludes with a discussion of the concept of the molecular assembler.

Molecular Assemblers

Drexler has proposed the molecular assembler, a device resembling an industrial robot which would be capable of holding and positioning reactive moieties in order to control the precise location at which chemical reactions take place. This general approach would allow the construction of large atomically precise objects by a sequence of precisely controlled chemical reactions. Much like the ribosome in biology, an assembler would build various classes of useful molecular structures following a sequence of instructions. During this process, the assembler would provide three-dimensional positional and full orientational control over the molecular component (analogous to the individual amino acid in the ribosome model) that is being added to a growing complex molecular structure (analogous to the growing polypeptide in the ribosomal model). In one approach, a molecular assembler may be capable of forming any one of several different kinds of chemical bonds (e.g., by changing tool tips), not just a single kind such as the peptide bond that the ribosome makes. In bonding atoms or molecules to one another, the assembler would provide any needed energy (especially if the reaction happens not to be energetically favoured) through physical force, thus performing mechanosynthesis (as opposed to the traditional means of chemical synthesis in solution). In another approach, a molecular assembler might be capable only of noncovalent assembly operations, wherein nanoparts are fabricated by other means and then presented to the assembler, which assembles the nanoparts into working nanomachines.

The first simple molecular assembler will almost certainly be a macroscale device, perhaps a modified SPM system as was being pursued by Zyvex in 1998. Multiple SPM heads could be equipped with a small number of nanoscale tool tips. In one scenario, nanoparts fabricated using bulk chemistry techniques would be inspected and selected by the SPM, then assembled one by one into working nanomachines (e.g., the desired useful nanoscale products). Such assembly operations will be very slow, because the placement of each new component may require simultaneous rotations and translations of large macroscale SPM components. Assembly time scales roughly linearly with assembler size because smaller assembler components moving at a given velocity need to travel less distance to accomplish a given physical operation, hence consuming less time and energy per physical operation. An important early developmental goal thus will be to design and fabricate nanoscale molecular assemblers.

At its most basic level, the simplest possible nanoscale molecular assembler may be comprised of one or more nanoscale manipulators. For example, the diamondoid telescoping manipulator arm described has about 4 million atoms excluding the base and control and power structures; doubling the size to account for support structures gives a molecular weight per arm of ~100 mega daltons and a total molecular volume of ~140,000 nm. Designed for high strength and stiffness, this robot arm should be able to hold a molecular fragment stiffly enough to make it react with a chosen end of a carbon-carbon double bond with an error rate of only 10^{-15}. The robot arm could guide chemical reactions with high reliability at room temperature in vacuo, with little need for sensing the positions of the molecules with which it is working. Alternatively, with appropriate tool tips, such an arm could grasp and manipulate individual prefabricated nanoparts in solution phase. Assembly of nanoparts without fabrication may require only a very small tool set.

The manipulator arm must be driven by a detailed sequence of control signals, just as the ribosome needs mRNA to guide its actions. However, such detailed control signals can be provided by external acoustic, electrical, or chemical signals that are received by the robot arm via an onboard sensor or power transducer, using a simple "broadcast architecture", a technique which can also be used to import power. Such transducers may be extremely small, on the order of (~10 nm) each. Interestingly, the biological cell may be regarded as an example of a broadcast architecture: the nucleus, located external to the cytoplasm, broadcasts mRNA chemical signals to millions of spatially diverse cytoplasmic ribosomes, thereby remotely controlling the construction of cellular proteins.

Thus a two-armed mechanical molecular assembler that receives power and instructions from some external agency may have a total molecular volume of ~300,000 nm (a cube ~67 nm on an edge), containing ~16 million atoms with a molecular weight of ~200 megadaltons—very roughly the mass and scale of a medium-size virus particle, such as an adenovirus. For comparison, the average enzyme (a biochemical crimping tool) weighs about ~0.1 megadalton, while a typical ribosome (a primitive "protein assembler") weighs ~4.2 megadaltons. A simple mechanical assembler in the 10-100 megadalton range cannot be ruled out. However, such a device might have a very small set of tool tips and an extremely limited manufacturing repertoire.

Multiple nanoscale assemblers each capable of independent simultaneous actuation may require control signals more conveniently provided by an onboard nanocomputer. This programmable nanocomputer must be able to accept stored instructions which are sequentially executed to direct the manipulator arm to place the correct moiety or nanopart in the desired position and orientation, thus giving precise control over the timing and locations of chemical reactions or assembly operations. The mechanical nanocomputer analyzed by Drexler requires ~16 nm per logic gate and ~40 nm per data register. If such components could be used to construct the equivalent of the most primitive 4-bit Intel 4004 microprocessor, then the nanocomputer could process ~10^5 bits/sec (~25,000 ops/sec) at a ~1 KHz clock speed within a mechanism volume of ~36,000 nm, again neglecting power supply, I/O linkages, and the like. An additional ~160,000 nm of rod logic registers adds ~1 kilobyte of onboard RAM memory or ~40 kilobytes of internal tape memory (conservatively assuming a tape storage density comparable to linear DNA). (A memory tape containing all bits needed for complete self-description may be considerably longer, even allowing for substantial data compression.) Doubling the total volume to account for support structure and other overhead gives a minimum mechanical nanocomputer molecular volume of ~400,000 nm, roughly 70 million atoms with a molecular weight of ~800 megadaltons. Thus the smallest nanocomputer-driven nanoscale molecular assembler with two manipulator arms may have a total molecular volume of ~700,000 nm (a cube ~88 nm on an edge), containing ~86 million atoms with a molecular weight of ~1 gigadalton.

By 1998, only a small amount of research targeted at actual assembler design had begun. Following Drexler's original discussions, during 1991-1998 R. Merkle authored or coauthored a continuing series of papers discussing various operational aspects and specific components of assembler design, including mechanosynthetic positional control, general design considerations for assemblers, the broadcast architecture, convergent assembly, binding sites, positioning devices, mechanosynthetic path sets, designs for a neon pump and a fine motion controller, and possible assembler casings. J.S. Hall has considered high-level designs for a nanoscale parts-fabrication and parts-assembly nanorobot. Zyvex, founded in 1996 by James von Ehr, has set itself the task of building the first programmable nanoassembler in a 5-10 year time frame (e.g., by ~2006). W. Goddard and colleagues have proposed a series of molecular dynamics simulations of simple assemblers, although by 1998 these studies evidently had not yet begun. According to Goddard's original proposal:

"Ultimately we need a programmable synthetic system to make a real device. Even though we may not have tools for all the chemical steps and may not have designs for all the pumps, engines, and transmissions needed, we propose to study the dynamics of simplified prototype assemblers. In these studies we anticipate having (1) a reservoir or supplies area for providing the various building units (atoms and fragments) required, (2) a work area in which we construct the nanomachine device (initially we will consider assembling tile structure on top of a diamond surface), and (3) a molecular scale nanohand which will extract the atoms from (1) and carry them to (2). We will then use extensions of our massive molecular dynamics program to operate the system: moving the tip from reservoir to work area, moving it to contact the appropriate surface site, moving it to regenerate the active tip, and then moving it back to add new atoms and molecules. This will include proper temperature effects, molecular vibrations, energy release upon the various chemical steps, etc. The ground rules here are that a realistic force field be used and that all pieces be treated at the atomic level (but some might be semirigid). This will use the force field developed for nanosynthesis. The purpose of these simulations is to examine issues of vibration caused by chemical forces as the tool picks up and delivers atoms to the growing surface. Also we want to consider the effect of energy release in the chemical steps on the thermal fluctuations in these systems (which may cause displacements and vibrations)."

Building mega-atom or giga-atom nanoproducts one at a time would be incredibly expensive and time consuming. For example, imagine that we wish to construct a simple medical nanorobot such as the 1-micron spherical respirocyte, which consists of ~18 billion atoms (dry structure). A factory employing a coordinated team of 100 macroscale SPM assemblers, each able to place ~1 atom/sec-SPM on a convergently-assembled workpiece achieves a pitiful manufacturing throughput rate of ~2 respirocytes per decade. Nanoscale assemblers with appendages ~10^6 times smaller than macroscale SPM assemblers might plausibly achieve net assembly rates of ~10^6 atoms/sec-nanoassembler, but even a production line employing ~300 such nanoassemblers (which the above mentioned 100-SPM factory team could build in ~1 year, assuming ~10^7 atoms/nanoassembler) can only manufacture ~1 respirocyte nanorobot per minute. At that rate, it would take ~2 million years to build the first ~1 cm therapeutic dosage containing ~10^{12} respirocyte nanorobots.

The necessary solution to this mass-production bottleneck is to employ any of several massively parallel approaches to manufacturing, including such techniques as self-assembly, convergent assembly, or, most usefully, self-replication. The basic advantage of self-replication is readily illustrated. Consider a single ~10^8 atom "seed" nano-assembler (having an onboard nanocomputer) that has been painstakingly built in ~10^6 sec using the above mentioned 100-SPM macroscale assembler team, working at ~1 atom/ sec-SPM. The seed nanoassembler is first programmed to build a copy of itself, which it accomplishes (working at ~10^6 atoms/sec-nanoassembler) in ~100 sec. These two nanoassemblers then each build a copy of themselves in another ~100 sec; now there are four nanoassemblers. After ~48 generations, requiring a total of ~80 minutes to complete, there are ~3 × 10^{14} nanoassemblers. These nanoassemblers are then reprogrammed for the manufacture of 18-billion-atom respirocytes and are fed the appropriate, presumably different, feedstock. This vastly expanded nanoassembler manufacturing system can now produce ~10^{12} respirocytes (~1 cm therapeutic dose) per minute.

Numerous complexities and limitations have been ignored here. For example, exponential growth cannot continue indefinitely in a finite environment with limited materials transport speeds—e.g. a 2-nm wide nanopart suspended in 310 K water diffuses only ~26 nm in 10^{-6} sec. Mechanical replicating systems designed purely for molecular manufacturing may be inflexible and brittle, employing limited energy resources and materials feedstocks not found outside of the immediate manufacturing environment.

The design of machines able to make copies of themselves was first described by von Neumann. Many variations on this theme have been reviewed by Freitas and Gilbreath and Sipper, and design for self-replication in the context of nanoscale assemblers has been considered by Drexler, Merkle, and Hall. As Drexler notes: "It may seem somehow paradoxical that a machine can contain all the instructions needed to make a copy of itself, including those selfsame complex instructions, but this is easily resolved. In the simplest approach, the machine reads the instructions twice: first as commands to be obeyed, and then as data to be copied. Adding more data does not increase the complexity of the data-copying process, hence the set of instructions can be made as complex as is necessary to specify the rest of the system. By the same token, the instructions transmitted in a replication cycle can specify the construction of an indefinitely large number of other artifacts."

The estimated information content of self-replicating systems—the length of the instruction tape—can be surprisingly small. Von Neumann's original analysis concluded that perhaps twelve different kinds of units of unknown complexity could be required as building materials, and Haldane inferred that as many as $\sim 10^5$ individual parts might be needed to make a replicator. This inference was refuted just three years later with the arrival of the first in a stream of very simple but ingenious designs for mechanical self-replicating machines that were assembled and operated in the late 1950s. In the first example, a pair of joined ratchet-like blocks, when placed in a "sea" of left and right blocks and then physically agitated, replicated itself from this well-ordered input substrate, making more block-pairs until all the single blocks were used up, a process similar to chemical autocatalysis.

More complex congeries of blocks, including clever four-block, eight-block, and 12-block replicators that could replicate themselves in a sea of blocks were presented by Penrose, Jacobson demonstrated a 3-unit replicator consisting of toy train engines circulating on HO model railroad tracks, and Morowitz designed a simple two-unit device with one unit comprised of about a dozen components including switches, batteries and electromagnets, that could assemble copies of itself from parts floating on the water surface of a bathtub. In 1998, Lohn and colleagues gave two designs for simple self-replicating systems that could be constructed out of wood, batteries and electromagnets, with explicit analogies to nanoscale fabrication. Several important conclusions may be drawn from these examples:

First, replication is fundamentally so simple a task that machines capable of displaying this behaviour predate most of the modern electronic computer era.

Second, assembly is an inherently simpler operation than fabrication. A complex part may embody hundreds or thousands of prior fabrication and assembly operations, yet may be installed within an assemblage in a single step. Hence nanopart assembly may be an easier candidate for early implementation in first-generation self-replicating molecular manufacturing systems than is molecular fabrication.

Third, and most important, the simplest nanoreplicator may require the ability to assemble, but not to (atomically) fabricate, in order to replicate itself. A nanomachine capable of assembly alone can replicate itself only from a very limited set of well-ordered input materials of relatively high complexity, but such a machine can be extremely simple both in structure and function. Certainly a nanomachine capable of both assembly and fabrication can replicate itself from a more diverse and disordered set of more elementary input materials but only at the expense of far greater internal structural and functional complexity.

At least two distinct models of replication have been identified in replicating systems design. The first model may be called the "unit replication" or organismic model, in which the replicator is an independent unit which employs the surrounding substrate to directly produce an identical copy of itself; both the original and the copy remain fertile and may replicate again, thus exponentiating their numbers. The second model may be called the "*unit growth*" or factory model, in which a population of specialist devices, each one individually incapable of self-replication, can collectively fabricate and assemble all necessary components comprising all specialist devices within the system, hence the factory is capable of expanding its size (or of manufacturing duplicate factory systems) indefinitely in an appropriate environment. The factory model is sometimes called "*bootstrapping*," which is any production of more productive capacity, from less. T. McKendree likens the situation to an ant colony: "There necessarily is the line of reproducing queens, making a sufficient number of ants feasible; [however,] most of the queen's products are 'workers' that perform useful functions, but themselves are incapable of reproduction."

It cannot be emphasized too strongly that mechanical medical nanodevices will not self-replicate inside the human body, nor will they have any need for self-replication themselves. Machines that perform medical tasks are fundamentally different from machines that manufacture other machines. As R. Merkle explains: "While self-replicating systems are the key to low cost [manufacture], there is no need (and little desire) to have such systems function in the outside world. Instead, in an artificial and controlled environment they can manufacture simpler and more rugged systems that can then be transferred to their final destination. Medical devices designed to operate in the human body don't have to self-replicate: we can manufacture them in a controlled environment and then inject them into the patient as needed. The resulting medical device will be simpler, smaller, more efficient and more precisely designed for the task at hand than a device designed to perform the same function and self-replicate. This conclusion should hold generally: optimize [product] device design for the desired function, manufacture the [product] device in an environment optimized for manufacturing, then transport the [product] device from the manufacturing environment to the environment for which it was designed. A single device able to do everything would be harder to design and less efficient."

In an effort to stimulate scientific and engineering interest in constructing the first nanoassembler, in November 1995 the Foresight Institute created the $250,000 Feynman Grand Prize, with funding contributed by Zyvex founder James von Ehr and St. Louis venture capitalist Marc Arnold. The prize will be awarded to the individual or group that first achieves both of two significant nanotechnology breakthroughs—first, the design and construction of a functional nanometer-scale robotic arm, and second, the design and construction of a functional 8-bit adder computing nanodevice.

According to Grand Prize rules, the robotic nanomanipulator must fit entirely inside a 100-nm cube, carry out actions directed by input signals of specified types, be able to move to a directed sequence of positions anywhere within a 50-nm cube, complete all directed actions with a positioning accuracy of 0.1 nanometer or better, and perform at least 1,000 accurate, nanometer-scale positioning motions per second for at least 60 consecutive seconds. The adder must fit entirely within a 50-nm cube, and must be capable of adding accurately any pair of 8-bit binary numbers, discarding overflow, accepting input signals of specified types, and producing its output as a pattern of raised nanometer-scale bumps on anatomically precise and level surface. (J.S. Hall notes that a conventional 8-bit adder may be constructed using a total of 94 AND, OR, and NOT gates; using XOR gates, the total may be reduced to 37 gates.) Both devices may accept inputs from acoustic, electrical, optical, diffusive chemical, or mechanical means, although any mechanical driving mechanism used for input must be limited to a single linkage that either slides or rotates on a single axis. To demonstrate the capacity for mass production, at least 32 copies of each device must be provided for analysis and destructive testing by judges. Once nanoassemblers are available, the design of far more complex nanomachines each containing tens or hundreds of billions of precisely arranged atoms will require new molecular CAD tools and techniques, including automated hierarchical design decomposition of objects (that must be built up from an array of nanoparts) and a shape description language.

Nanosystem design compilers, conceptually similar to silicon compilers that generate a complex pattern of transistors and conductors from an abstract specification of the properties of a digital circuit, will also be required for efficient nanorobot design. Given a fully specified design that meets all constraints necessary to physically permit assembly, all operations of the assembly process must be specified using an assembly process compiler. Explains Drexler: "The design of structures and assembly procedures by hierarchical decomposition directly generates a tree of assembly steps. If this tree has been chosen to generate parts of the appropriate sizes and numbers, then it can be mapped onto a manufacturing layout. The vast number of manipulator motions to be specified at the finest, earliest, and most-dispersed levels of the assembly process can be planned in parallel by identical software systems running almost independently on separate processors. The result of this parallel assembly process compilation is a set of instructions that, when executed by the transportation system and manipulator controllers of a manufacturing mechanism, will result in the assembly of an object corresponding to the initial design." Scale-up to larger systems, including the control, coordination, and programming of large aggregates of cooperating nanorobots.

Since molecular manufacturing systems can be used to make more molecular manufacturing systems, it is believed that the capital costs of production can be quite low. Drexler claims that an analysis of inputs, outputs, and productivity suggests the total cost of production can be in the range familiar in agriculture and in the production of industrial chemicals—on the order of tens of cents per pound. Merkle has made similar assertions, for example: "Eventually (after amortization of possibly quite high development costs), the price of assemblers (and of the objects they build) should be no higher than the price of other complex structures made by self-replicating systems. Potatoes—which have a staggering design complexity involving tens of thousands of different genes and different proteins directed by many megabits of genetic information—cost well under a dollar per pound." Such analogies from conventional agriculture or industrial chemistry are not strictly applicable to the medical products and pharmaceutical areas, where structural precision, product repeatability, and intensive quality control are of supreme importance, and where most products are subject to rigorous government regulation—unlike potatoes, which vary widely in size, shape, and molecular constitution, and generally do not require stringent product liability insurance, multiphase clinical trials, or FDA approval.

Still, it is likely that molecular manufacturing will eventually allow reasonably inexpensive production of large batches of designed medical nanorobots. As a very crude analogy, even in 1995 a typical mail-order bioscience vendor could manufacture 15 micromoles (~9 $\times 10^{18}$ product items) of customized oligonucleotides for \$18/base up to at least 110 bases with a convenient 48-hour turnaround on orders; other online vendors offer similar or slightly higher prices up to 200-mer sequences with additional charges for necessary purifications. If a billion-base structure could be induced to self-assemble from 10 million different sets of 100-mer custom-built oligonucleotides (total ~20 billion atoms), forming a desired ~20-billion-atom nanomachine, then, optimistically assuming a ~100% yield, the cost would be ~\$2 $\times 10^{9}$ per nanomachine or ~\$2000 per ~1 cm of product volume which would include 10^{12} finished nanomachines. In 1998, many specialty biotechnology-related drug treatments were comparably expensive. For example, treatments for multiple sclerosis using two closely related ~22.5 kilodalton interferons were administered in dosages of ~1 milligram/week at a treatment cost of ~\$10,000/yr, which is equivalent to a treatment cost of ~\$2 $\times 10^{9}$ (2 "nanodollars") per giga-atom. The famous HeLa (Henrietta Lacks) cell line is an example of a once tiny population of unique cancerous cells that were purposely replicated in vitro (~24-hour replication time) over many decades and are now widely available worldwide at very low cost (e.g., \$216 per 1-cm ampule of the ATCC CCL-2 HeLa cell line). Swallowable therapeutic pills containing bacteria (i.e., natural biological nanomachines) are already widely available over the counter for gastrointestinal refloration, as for example Salivarex which "contains a minimum of ~2.9 billion beneficial bacteria per capsule", and Alkadophilus which "contains 1.5 billion organisms per capsule", both at a 1998 price of ~\$0.2 $\times 10^{9}$ per bacterium.

Another useful analogic approach to cost estimation is suggested by the plunging price of computer power during the 20th century. This trend may give us some confidence that even if the first individual 1 micron nanorobot costs ~\$100,000 to assemble (much like the first Nippondenso microcar; eventually the unit price may fall at least 14 orders of magnitude to below ~\$$10^9$ per nanorobot or <\$1000 per ~1 cm dosage (10^{12} nanorobots). Recycling or remanufacturing each nanomachine ~1000 times before final discard would then imply a net ~\$1/cm treatment cost. Of course, in 1998 all such estimates were mere guesswork, but the ultimate expectation of relatively low-cost nanomedical treatments was certainly rational, if not provable.

This Chapter has demonstrated that there are many different pathways leading toward molecular nanotechnology, each one of which provides incremental benefits to motivate further travel down the paths. The likelihood of all the paths being blocked is low, even if we have little confidence that we can predict which approach will succeed first. Progress along the many pathways will provide precursor products, including some early medical applications, but the ultimate achievement of molecular manufacturing will finally make nanomedicine feasible.

As Drexler concluded Nanosystems: "Each step along will present great practical challenges, but each step will also bring valuable new capabilities. The long-term rewards, measured in terms of scientific and technological capabilities, appear large." This author agrees. For the remainder of the present work, we shall assume that a molecular manufacturing technology will someday exist that can economically manufacture macroscopic batches of microscopic machines comprised of atomically-precise nanoscale components.

6

Web of Information

To prepare for the assembler breakthrough, society must learn to learn faster. Fact forums will help, but new technologies may help even more. With them, we will be able to spread, refine, and combine our information far faster than ever before. *Information* overload has become a well-known problem: pieces of knowledge pile up too fast for people to sort and make sense of them. Thousands of technical journals cover thousands of subjects. Published articles mount by over a million per year. Fact forums help us to clear away falsehoods, which will smooth our efforts to make sense of the world. But any such formal institution will be overwhelmed by the modern flood of information: fact forums will be able to deal with only a fraction—though an important fraction—of the facts, and they will inevitably be somewhat sluggish. Formal institutions can tap only a tiny fraction of the mental energies of our society.

Today, our information systems hamper our progress. To see the problem, imagine handling a piece of information: You uncovered it—how do you spread it? Someone else published it—how do you find it? You found it—where do you file it? You see an error—how do you correct it? Your file grows—how do you organize it? We now handle information clumsily. Our traditional electronic media are vivid and entertaining, but they are ill suited to handling complex, long-term debates; how could you, as a viewer, file, organize, or correct the information in a television documentary? In short, how could you make it a well-integrated part of an evolving body of knowledge? We can handle complex debates better using paper media, yet the weeks (or years) of delay in a typical publication process slow debate to a crawl.

And even paper publications are difficult to file, organize, and correct. Printers produce bundles of inked paper; through heroic efforts, librarians and scholars manage to link and organize them in a loose fashion. Yet indexes, references, and corrections simply add more pages or more editions, and tracing the links they represent remains tedious. Books and other paper bundles work, after a fashion. They hold many of our cultural treasures, and we now have no better way to publish most things. Still, they leave great room for improvement.

Our trouble in spreading, correcting, and organizing information leaves our shared knowledge relatively scarce, incorrect, and disorganized. Because established knowledge is often hard to find, we often do without it, making us seem more ignorant than we need be. Can new technologies help us?

They have before. The invention of the printing press brought great advances; computer-based text services promise yet more. To see how our information systems could be better, though, it may help to see how they could be worse. Consider, then, an imaginary mess and an imaginary solution:

Evolving Techniques

Once upon a time, there lived a people with an information problem. Though they had replaced their bulky clay tablets with paper, they used it oddly. In the heart of their land stood a stately dome. Beneath the dome lay their great Chamber of Writings. Within this chamber lay a broad mound of paper scraps, each the size of a child's hand.

From time to time, a scholar would journey to this temple of learning to offer knowledge. A council of scribes would judge its worth. If it proved worthy, they would inscribe it on a scrap of paper and ceremoniously fling it upon the heap.

From time to time, some industrious scholar would come to seek knowledge—to rummage through the heap in search of the needed scrap. Some, skilled in such research, could find a particular scrap in as little as a month. The scribes always welcomed researchers: they were so rare. We moderns can see their problem: in a disorderly heap, each added scrap buries the rest (as on so many desks). Every scrap is separate, unrelated to the others, and adding references would provide little help when finding a scrap takes months. If we used such a heap to store information, our massive, detailed writings on science and technology would become almost useless. Searches would take years, or lifetimes.

We moderns have a simple solution: we place pages in order. We place page after page to make a book, book after book to fill a shelf, then fill a building with shelves to make a library. With pages in order, we can find them and follow references more rapidly. If the scribes employed scholars to stack scraps by subject, their research would grow easier.

Yet, when faced with stacks on history, geography, and medicine, where should the scholars put scraps on historical geography, geographical epidemiology, and medical history? Where should they put scraps on "The History of the Spread of the Great Plague"? But in our imaginary land, the scribes choose another solution: they send for a magician. But first they turn scholars loose in the chamber with needles and thread to run strands from scrap to scrap. Thread of one colour links a scrap to the next in a series, another colour leads to a reference, another to a critical note, and so forth. The scholars weave a network of relationships, represented by a network of threads. At last, the magician (with flashing eyes and floating hair) chants a spell, and the whole mess heaves slowly into the air to float like a cloud in the dome. Ever after, a scholar holding a scrap need only tap a thread knotted to its edge to make the linked scrap leap to hand. And the threads, magically, never tangle.

Now the scholars can link scraps on "The History of the Spread of the *Great Plague*" to related scraps on history, geography, and medicine. They can add all the notes and texts they please, linking them to best advantage. They can add special index scraps, able to bring instantly to hand whatever they list. They can place links wherever they wish, weaving a network of knowledge to match the connections of the real world. We, with our inert stacks of paper, could only envy them—if we didn't have computers.

Computer Technology

In 1945, Vannevar Bush proposed a system he called a "*memex.*" It was to be a desk-sized device, crammed with microfilm and mechanisms, able to display stored pages and let the user note relationships among them. A microfilm memex was never built, but the dream lived on. Today, computers and screens are becoming cheap enough to use for ordinary reading and writing. Some paper publishers have become electronic publishers, making magazines, newspapers, and journals available through computer networks. And with the right programs, texthandling computers will let us link this information in ways even better than magic thread.

Theodor Nelson, the originator of the idea, has dubbed the result "*hypertext*": text linked in many directions, not just in a one-dimensional sequence. Readers, authors, and editors using a hypertext system will generally ignore the workings of its computers and screens just as they have generally ignored the mechanics of photocomposition and offset lithography in the past. A *hypertext system* will simply act like magic paper; anyone who fiddles with it will soon become familiar with its basic abilities. Still, a description of one system's structure will help in explaining how hypertext will work.

In the approach followed by the Xanadu hypertext group, the core of the system—the back end—is a *computer network* able to store both documents and links between documents. An initial system might be a single-user desktop machine; eventually a growing network of machines will be able to serve as an electronic library. Stored documents will be able to represent almost anything, whether novels, diagrams, textbooks, or programs—eventually, even music or movies.

Users will be able to link any part of any document to any other. When a reader points to one end of a link (whether it is shown on the screen by underlining, an asterisk, or a picture of coloured thread), the system will fetch and display the material at the other end. Further, it will record new versions of a large document without storing additional copies; it need store only the parts that are changed. This will let it inexpensively store the earlier versions of any document published and modified on the system. It will do all this rapidly, even when the total amount of information stored becomes immense. A network of such machines could eventually mature into a world electronic library.

To locate material in most computer-based text systems, the user must supply key words or obscure codes. Hypertext, too, will be able to link text to codes, to key words, or even to a simulated card catalog, but most readers will probably prefer just to read and point to links. As Theodor Nelson has remarked, hypertext will be "a new form of reading and writing, in a way just like the old, with quotations and marginalia and citations. Yet it will also be socially self-constructing into a vast new traversable framework, a new literature. What the reader will see when browsing through this framework will depend partly on the reader's own part of the system, the "*frontend*" machine, perhaps a personal computer. The back end will just file and fetch documents; the front end will order them fetched at the reader's request and will display them to suit the reader's taste.

To imagine how this will appear to a user, picture a screen the size of this open book, covered with print the size you are reading now-clear print, on a good screen. Today the screen would resemble a television set, but within a few years it could be a booklike, lap-sized object with a cord to an information outlet.

By keeping track of links (say, between outlines, drafts, and reference materials), *hypertext* will help people write and edit more ambitious works. Using hypertext links, we can weave our knowledge into coherent wholes. John Muir observed that "When we try to pick out anything by itself, we find it hitched to everything else in the universe." Hypertext will help us keep ideas hitched together in ways that better represent reality. With hypertext, we will be better able to gather and organize knowledge, increasing our effective intelligence. But for information gathering to be effective, it must be decentralized; information scattered among many minds cannot easily be put into the system by a few specialists. The Xanadu group proposes a simple solution: let everyone write, and have the system automatically pay royalties to the authors whenever readers use their material. Publishing will be easy and people will be rewarded for providing what other people want.

Imagine what you yourself have wanted to say about ideas and events. Imagine the insightful comments that are even now fading from the memories of both speakers and listeners all over the world. On a *hypertext system*, comments will be easy to publish and easy to find. Imagine the questions that have bothered you. You could publish them, too; someone finding an answer could then publish a reply.

Since everyone on the system will be able to write text and links, the hypertext network will accumulate great stores of knowledge and wisdom and even greater heaps of utter garbage. Hypertext will include old news, advertising, graffiti, ranting, and lies—so how will a reader be able to avoid the bad and focus on the good? We could appoint a central editorial committee, but this would destroy the openness of the system. Sorting information is itself an information problem for which hypertext fortunately will help us evolve good solutions. Since hypertext will be able to do almost everything that paper systems can, we can at least use the solutions we already have. Publishers have established reputations in the paper-text media, and many of them have begun to move into electronic publishing. On a hypertext system they will be able to publish on-line documents that meet their established standards. Readers so inclined will be able to set their front-end systems to display only these documents, automatically ignoring the new garbage. To them, the hypertext system will seem to contain only material by established publishers, but material made more available by electronic distribution and by hypertext links and indexes. True garbage will still be there (so long as its authors have paid a small storage fee for their material), yet garbage need not intrude on any reader's screen.

But we will be able to do better than this. Approval of a document (shown by links and recommendations) can come from anyone; readers will pay attention to material recommended by whomever they respect. Conversely, readers who find documents they like will be able to see who has recommended them; this will lead readers to discover people who share their interests and concerns. Indirectly, hypertext will link people and speed the growth of communities.

When publishing becomes so fast and easy, writers will produce more material. Since hypertext will encourage free-lance editing, editors will find themselves with more work to do. Documents that quote, list, and link other documents will serve as anthologies, journals, or instant-access indexes. The *incentive of royalties* will encourage people to help readers find what they want. Competing guides to the literature will swiftly appear—and guides to the guides.

Hypertext links will be better than paper references, and not merely in speed. Paper references let an industrious reader follow links from one document to another—but try finding which documents refer to one you are reading! Today, finding such references requires the cumbersome apparatus of a citation index, available only in research libraries, covering limited topics, and months out of date. Hypertext links will work in both directions, letting readers find comments on what they are reading. This means a breakthrough: it will subject ideas to more thorough criticism, making them evolve faster. The evolution of knowledge—whether in philosophy, politics, science, or engineering—requires the generation, spread, and testing of memes. Hypertext will speed this process. Paper media handle the process of generation and spread fairly well, but they handle testing clumsily.

Once a bad idea reaches print, it takes on a life of its own, and even its author can seldom drive a stake through its heart. A devastating refutation of the bad idea becomes just another publication, another scrap of paper. Days or years later, readers who encounter the bogus idea will still be unlikely to have chanced upon its refutation. Thus, nonsense lives on and on. Only with the advent of hypertext will critics be able to plant their barbs firmly in the meat of their targets. Only with hypertext will authors be able to retract their errors, not by burning all the libraries or by mounting a massive publicity campaign, but by revising their text and labeling the old version "retracted." Authors will be able to eat their words quietly; this will give them some compensation for the fiercer criticism.

Critics will use clear refutations to skewer nonsense (such as false limits to growth), clearing it from the intellectual arena - though not from the record - almost as soon as it pops into sight. Guides to good criticism will help readers see whether an idea has survived the worst objections yet raised. Today, the absence of known criticism doesn't mean much, because brief critical comments are hard to publish and hard to find. In an established hypertext system, though, ideas that have survived all known criticism will have survived a real challenge. They will gain real and growing credibility.

Advantages of Hypertext

The advantages of hypertext run deep; this is why they will be great. Hypertext will let us represent knowledge in a more natural way. Human knowledge forms an unbroken web, and human problems sprawl across the fuzzy boundaries between fields. Neat rows of books do a poor job of representing the structure of our knowledge. Librarians have laboured to make these rows more like nets by inventing better ways to index, reference, and arrange pieces of paper. Yet despite the noble efforts and victories of librarians, library research still daunts all but a dedicated minority of the reading public. Libraries have evolved toward hypertext, yet the mechanics of paper still hobbles them. Hypertext systems will let us take a giant step in a direction we have been moving since the invention of writing.

Our very memories work through associations, through links that make recollections recallable. AI workers also find associations essential to making knowledge useful; they program what they call "*semantic nets*" to build knowledge representation systems. On paper, associations among words make a thesaurus useful; in the mind, one's working vocabulary relies on fast, flexible associations among words. Indeed, relationships in memory supply the context that gives meaning to our ideas. Using *hypertext*, people will associate ideas through published links, enriching their meaning and making them more available - indeed, making them more like parts of our own minds. When we change our minds about what the world is like and where it is going, we change our internal networks of knowledge. Reasoned change often requires that we compare competing patterns of ideas.

To judge a worldview presented in a book, a reader must often remember or reread explanations from earlier pages—or from a conflicting article seen last year. But human memory is faulty, and digging around in old paper often seems like too much work. Knowing of this problem, authors vacillate between putting too much in (thus boring their readers) and leaving too much out (thus leaving weak spots in their discussions). Inevitably, they do both at once.

Hypertext readers will be able to see whether linked sources support an idea or linked criticisms explode it. Authors will write pithy, exciting summaries of ideas and link them to the lengthy, boring explanations. As authors expound and critics argue, they will lay out their competing worldview networks in parallel, point by point. Readers still won't be able to judge ideas instantly or perfectly, but they will be able to judge them faster and better. In this way, hypertext will help us with a great task of our time: judging what lies ahead, and adjusting our thinking to prospects that shake the foundations of established worldviews. Hypertext will strengthen our foresight.

By now, many useful applications of a mature *hypertext system* will be obvious—or as obvious as they can be today, before we have experienced them directly. Carrying news is one such obvious application. News shapes our view of the world, but modern media sharply limit what reporters can portray. Often, stories about technology and world events only make sense in a broader context, but the limited space and onrushing deadlines of publishing strip needed context from stories. This weakens our grasp of events. Using hypertext, reporters will find it easy to link today's news to broader background discussions. What is more, the people in the stories and casual observers will be able to have their say, linking their comments to the reporter's story.

Advertising greases the wheels of the economy, leading (and misleading) us to available products. Well-informed consumers can avoid shoddy, overpriced goods, but the needed research and comparison shopping gobbles time. On a hypertext system, though, consumer service companies will assemble comparative catalogs, linking descriptions of competing products to one another, to test results, and to reports from consumers.

In education, we learn best when we are interested in what we read. But most books present ideas in just one sequence, at just one level of difficulty, regardless of a learner's background or interests. Again, popular demand will favour the growth of useful networks in hypertext. People will make links between similar presentations written at different levels. Students will be able to read at a comfortable level, peeking at parallel discussions that reach a bit deeper. Hard material will grow easier to handle, because links to primers and basic definitions will let readers pause for review—instantly, privately, and without embarrassment. Other links will lead in all directions to related material; links in a description of a coral reef will lead to both texts on reef ecology and tales of hungry sharks. When we can gratify momentary interests almost instantly, learning will become more fun. More people might then find it addictive.

Due process will thrive in hypertext. Because it will be open to all sides, and will allow questioning, response, and so forth, hypertext debate will have an inherent dueprocess quality. Indeed, hypertext will be an ideal medium for conducting fact forums. Forum procedures, in turn, will complement hypertext by distilling its wideranging debates into a clear (though tentative) statement of key technical facts.

In a final obvious effect, hypertext will reduce the problem of out-of-context quotes: readers will be able to make the original context reappear around any quote in the system at the touch of a button. This will be valuable, and not just to prevent misrepresentation of an author's position; indirect benefits may matter more. Reasonable statements torn loose from their background can seem absurd, but hypertext authors will know that "*absurd*" quotations will lead readers directly back to the author's original context. This will encourage bolder writing, giving memes based on evidence and reason an advantage over those based on mere convention and timidity.

Perhaps the most important benefit of hypertext will be a new ability to see absences. To survive the coming years, we must evaluate complex ideas correctly, and this requires judging whether an argument is full of holes. But today we have trouble seeing holes. Still more difficult is recognizing the absence of fatal holes, yet this is the key to recognizing a sound argument. Hypertext will help us. Readers will scrutinize important arguments, attaching conspicuous objections where they find holes. These objections will make holes so consistently visible that an absence of good objections will clearly indicate an absence of known holes. It may be hard to appreciate how important this will he: the human mind tends not to recognize the problems caused by our inability to see the absence of holes, to say nothing of the opportunities this inability makes us miss.

For example, imagine that you have an idea and are trying to decide whether it is sound and worth publishing. If the idea isn't obvious, you might doubt its truth and not publish it. But if it does seem obvious, you might well assume that it has already been published, but that you just can't find out where. Hypertext, by making things much easier to find, will make it easier to see that something has not been published. By making holes in our knowledge more visible, hypertext will encourage holefilling. To understand and guide technology, we need to find the errors-including omissions—in complex technological proposals. Because we do this poorly, we make many mistakes, and the visibility of these mistakes makes our incompetence a vivid and menacing fact. This encourages prudence, yet it can also encourage paralysis: because we have difficulty seeing holes, we fear them everywhere, even where they do not exist. Hypertext will help build confidence, where confidence is justified, by exposing problems more reliably.

Dangers of Hypertext

Like most useful tools, hypertext could be used to do harm. Though it will help us keep track of facts, it could also help governments keep track of us. Yet, on balance, it may serve liberty. Designed for decentralization—with many machines, many writers, many editors—*hypertext* may help citizens more than it helps those who would rule them. Government data banks are growing anyway. *Hypertext systems* might even help us keep an eye on them.

Relying on electronic publishing holds another danger. Governments in the United States and elsewhere have often interpreted the ideal once expressed as "*freedom of speech*" and "of the press" to mean only freedom to talk and to sell inked paper. Governments have regulated the use of radio and television, requiring them to serve a bureaucracy's shifting notion of the public interest. Practical limitations on the number of broadcasting channels once gave some excuse for this, but those excuses must stop here. We must extend the principles of free speech to new media. We would be horrified if the government ordered agents into libraries to burn books. We should be equally horrified when the government seeks to erase public documents from electronic libraries. If hypertext is to carry our traditions, then what is published must remain so. An electronic library will be no less a library for its lack of shelves and paper. Erasure will make no flame and smoke, but the stench of book-burning will remain..

Searching through Internet

Some of the benefits already described will only result from a large, highly evolved hypertext system—one already serving as a forum for broad debates and on its way to becoming a world *electronic library*. Such a system may not have time to mature before the assembler and AI breakthroughs arrive. For hypertext to get off the ground, small systems must have practical applications, and for hypertext to help us handle the technology race, small systems must have an effect beyond their size. Fortunately, we can expect substantial benefits almost from the outset.

Individual *hypertext machines* will be able to serve several users at once. Even without linking to anything in the outside world, they will help companies, associations, and research groups handle complex information. Outside links will be easy, though. The number of publicly available data bases has grown from a few dozen in the mid-1960s to a few hundred in the mid-1970s to a few thousand in the mid-1980s. Companies have made these available through *computer networks*. Hypertext systems will be able to fetch material from these data bases, storing the access codes instead of the actual text. This information will only seem to be in the hypertext system, but that will be good enough for many purposes. People will use early systems to provide a community with a dialup service like that of existing computer bulletin boards but better. Special-interest discussion groups have already emerged on computer networks; they will find hypertext a better medium for exchanging information and views.

Early hypertext systems will also help us build and run organizations. Ordinary *computer conferencing* (simply sending short messages back and forth) already helps groups communicate. Advantages over face-to-face conferences include lower costs (no need to travel), smoother interactions (no need to wait or to interrupt people), and a better meeting of minds (through clearer messages and fewer personality clashes). *Hypertext communications* will extend these benefits by giving participants better tools for referencing, comparing, and summarizing. Because hypertext debates will need no single editor, they will allow organizations to become more open.

Using a hypertext service by telephone from a home computer during off-peak hours will probably cost several dollars per hour at first. This cost will fall over time. For several decades now, the real cost of computers has dropped by about a factor of ten every ten years; the cost of communications has also declined. Hypertext systems will be affordable to a substantial number of people almost as soon as they become available. Within ten years, the costs seem likely to fall low enough for mass-market use.

Electronic publishing is already catching on. The Academic American Encyclopedia, structured as a simple hypertext, has been made available to 90,000 subscribers, including 200 libraries and eight schools. Time magazine reports that children use it eagerly. Terminals in libraries already can access the text of scores of newspapers, magazines, and professional journals. We won't need to wait for a universal system to enjoy universal benefits, because hypertext will begin to make a difference very early. We need hypertext in the hands of students, writers, researchers, and managers for the same reason that we need textbooks in schools, instruments in laboratories, and tools in workshops. Some books have made a great difference, even when read by fewer persons than one in a thousand, because they have sent new ideas rippling across society. Hypertext will do likewise, helping to refine ideas that will then spread more widely through the established print and broadcast media.

Hypertext Revolution

How will the *hypertext revolution* compare to the Gutenberg revolution? Some numbers suggest the answer. Printing with movable type cut the cost of books dramatically. In the fourteenth century, the King's Advocate of France had only seventy-six books, yet this was considered a large library. Books embodied weeks of skilled labour—copyists were literate. The peasant masses could neither afford books nor read them. Today, a year's labour can pay for thousands of books. Many homes hold hundreds; large libraries hold millions. Printing cut the cost of books by a hundredfold or more, setting the stage for mass literacy, mass education, and the ongoing world revolution of technology and democracy. And hypertext? Gutenberg showed Europe how to arrange metal type to print pages; *hypertext* will let us rearrange stored text and send it cross-country at the speed of light. Printing put stacks of books in the home and mountains of books in libraries; hypertext will in effect bring these mountains of books to every terminal. Hypertext will extend the Gutenberg revolution by increasing the quantity of information available.

Yet its other advantages seem greater. Today, following a reference in a library typically takes minutes; with luck, a few hundred seconds - but it can take days or more, if the material is unpopular and hence absent, or too popular and hence missing. Hypertext will cut this delay from hundreds of seconds to about one second. Thus where the Gutenberg revolution reduced the labour cost of producing text by several hundredfold, the hypertext revolution will reduce the labour cost of finding text by several hundredfold. This will be a revolution indeed. Making links more convenient will change the texture of text, bringing a revolution not merely in quantity but in quality. This increase in quality will take many forms. Better indexes will make information easy to find. Better critical discussion will weed out nonsense and help sound ideas thrive. Better presentation of wholes will highlight the holes in our knowledge. With abundant, available, high-quality information, we will seem more intelligent. And this will increase our chances of handling the coming breakthroughs right. What could be more useful? Next time you see a lie being spread or a bad decision being made out of sheer ignorance, pause, and think of hypertext.

7

THRESHOLD OF NANOTECHNOLOGY

In this chapter, we outline how emerging technologies can lead to nanotechnology. The actual path to nanotechnology—the one that history books will record—could emerge from any one of the research directions in physics, biochemistry, and chemistry has already been recounted, or (more likely) from a combination of them. The availability of so many good options builds confidence that the goal can be reached, even while it decreases confidence that some particular path will be fastest. To see how advances might cross the gap from present technology to early nanotechnology, let's follow one path out of the many possible.

BRIDGING THE GAP

One way to bridge the gap would through the development of an *AFM*-based *molecular manipulator* capable of doing primitive *molecular manufacturing*. This device would combine a simple molecular device—a molecular gripper—with an AFM positioning mechanism. An AFM can move its tip with precision; a molecular manipulator would add a gripper to the tip to hold a molecular tool. A molecular manipulator of this kind would guide chemical reactions by positioning *molecules*, like a slow, simple, but enormous *assembler*. (In our standard simulation view, where a molecular assembler arm fits in a room, the AFM apparatus of a molecular manipulator would be the size of a moon.) Despite its limits, an AFM molecular manipulator will be a striking advance.

Developing a Molecular Manipulator

Several years ago, researchers at the University of Brobdingnag began work on developing a molecular manipulator. To reach this goal, a team of a dozen physicists, chemists, and protein researchers banded together (some working full time, some part time) and began the creative teamwork needed to solve the basic problems.

First they needed to attach a gripper to an AFM tip. As grippers, they chose fragments of antibody molecules, the selectively sticky proteins that the immune system uses to bind and identify germs. If they could get the "back" of the molecule stuck onto a tip, then the "front" could bind and hold molecular tools. (The advantage of antibody fragments was this: freedom of tool choice. Since the late 1980s, researchers had been able to generate antibodies able to bind almost any preselected molecule-or molecular tool.) In parallel, the U. Brob AFM researchers worked on placing tips in a precise location and then holding them there with atomic accuracy for seconds at a time. This proved straightforward. They used techniques developed elsewhere during the early 1990s, adding only modest refinements.

They now had their gripper and a way of putting it where they wanted it, but they needed a set of tools. The gripper was like the chuck of a drill, waiting to have different bits fitted into its tool-

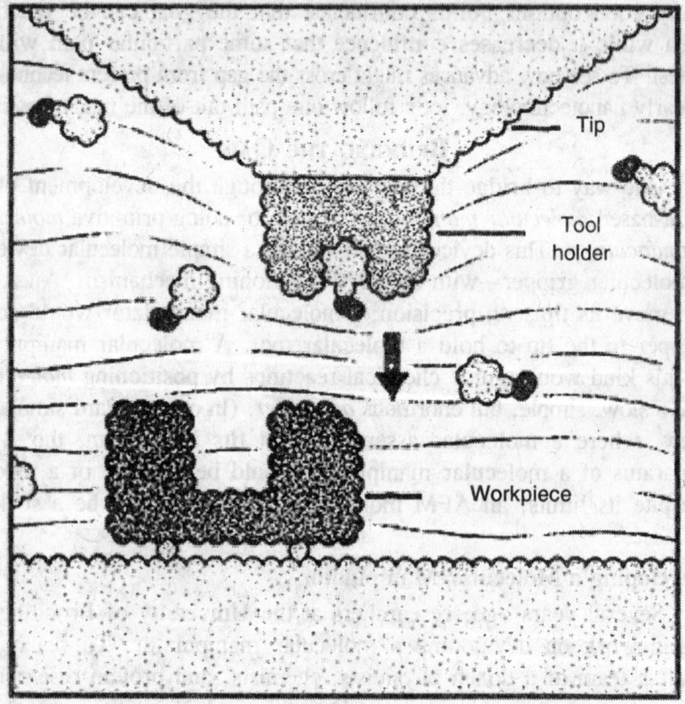

Fig. 7.1. Molecular manipulator.

holder slot. So as the final step, the synthetic chemists on the team made a dozen different molecular tools, all identical at one end but different at the other. The similar parts all bound to the same antibody tool-holder, slotting neatly into position. The different parts were all chemically reactive in different ways. Like the molecular tools in the hall of assembler arms, each of these tools could use a chemical reaction to transfer some *atoms* to a molecular object under construction.

Developing the molecular tool kit was the toughest part of the project; it took about as much work as had gone into duplicating the palytoxin molecule back in the 1980s. None of the tasks in the project demanded the solution of a deep scientific puzzle, and none demanded the solution of a notoriously difficult engineering problem. Each task had many possible solutions, the problem was to find a compatible set of solutions and apply them. After a few years, the solutions came together and the U. Brob research team began building new molecules by molecular manipulation. Now many teams are doing likewise.

Building with Molecular Grippers and Tools

To build something with the U. Brob team's AFM-based molecular manipulator system, you use it as follows: First, choose a surface to build on and place it under the tip in a pool of liquid. Then dunk the AFM tip into the liquid, bringing it down to the surface, and back it off a little. Construction can now begin as soon as a tool is loaded into the gripper.

Tubes and pumps can flow different liquids over the surface and past the gripper, carrying different tool molecules. If you want to do something with a tool of Type A, you wash in the proper liquid, and a Type A molecule promptly sticks the to the gripper. Once it is in the gripper, you can use the AFM mechanism to move it around and put it where you want it. Move it up to the surface at a convenient spot, wait a few seconds, and it reacts, forming a bond and leaving a molecular fragment attached to the spot you chose. To add a different fragment, you can use a tool of Type B: you back up the tip, flow in a fresh liquid carrying the new tools, and in a moment a tool of the new type is bound in place and ready to apply, either on or alongside the first spot. Step by step, you build up a precise molecular structure.

Each step takes only seconds. Molecular tools pop into the gripper in a fraction of a second, and used tools pop off at the same rate. Once the tip has positioned a molecule, it reacts quickly, about a million times faster than unwanted reactions at other sites. In this way, the molecular manipulator gives good control of where reactions will occur (though it is not as reliable as an advanced assembler would be). It is fairly fast by a chemist's standards—per cycle—but still a million times slower than an advanced assembler. It can perform a variety of steps, but isn't as flexible and capable as an advanced assembler. In short, it is hardly the last word in nanotechnology, yet is a great advance over what has gone before.

Products

With its ability to accelerate desired reactions by a factor of a million or so, the U. Brob team's molecular manipulator can perform 10,000 to 100,000 steps with good reliability. Back in the 1980s, chemists making protein molecules struggled to perform just one hundred steps. The U. Brob research team can now build structures that are stronger and easier to design than proteins: not floppy, folded chains, but rugged objects held together by a sturdy network of bonds. Though not as strong and dense as diamond, these structures are like bits of a tough engineering plastic. A specially adapted computer-aided design system makes it easy to design molecular objects made from these materials.

Yet the AFM-based molecular manipulator has one grave disadvantage: It does chemistry one molecule at a time, and it ties up a machine as expensive as a car for hours or days to produce that one large molecule. Some molecules, though, are valuable enough to be worth building even one at a time. These draw prompt attention.

A single molecule isn't much use as a dye, a drug, or a floor wax, but it can have substantial value if it provides useful information. The U. Brob team quickly publishes a pile of scientific papers based on experiments with single molecules: they build a molecule, probe it, report the results, and build another. Some of these results show chemists elsewhere in the multibillion-dollar chemical industry how to design new catalysts, molecules that can help make other molecules more cheaply, cleanly, and efficiently. This information is worth a lot. Three new products of special interest are among the first to be made.

Molecular electronics

Molecular electronics begins with experiments conducted by a research group at a computer chip company. They use their molecular manipulator to build single molecules and probe them, gradually learning how to build the parts needed for molecular electronic computers. These new computers don't immediately become practical, because the costs are too high for making such large molecules with AFM-based technology. Yet some companies begin to produce simpler molecular electronic devices for use in sensors and specialized high-speed signal processing. A specialty industry is born and begins to expand.

Gene reader

The gene reader is a complex molecular device built on the surface of a chip. The biologists who built the reader combined proteins borrowed from *cells* with special-purpose *molecular machines* designed from scratch. The result was a molecular system that binds *DNA* molecules and pulls them past a read-head-like tape through a tape recorder. The device works as fast as some naturally occurring molecular machines that read DNA, with one key advantage: it outputs its data electronically. At that speed, a single device can read a human genome in about a year. Though still too expensive for a doctor's office, these readers are promptly in great demand from research laboratories. Another small industry is born.

The third product is far more important, in the long run: replacement tips for molecular manipulators, grippers, and tools that are better than the originals. With these new, more versatile devices, researchers are now building more ambitious products and tools.

More to Nanotechnology

While the physicist-led team at U. Brob was finishing its work on the AFM-based molecular manipulator, a chemist-led team at the University of Lilliput was working furiously. They saw the U. Brob desktop machine as too large and its expected products as too expensive. Even back in the 1980s, David Biegelsen of the Xerox Palo Alto Research Center had noted, "The main drawback I see to using a hybrid protoassembler [AFM-based molecular manipulator] is that it would take a long time to build just one unit. Building requires a series of atom-by-atom construction steps. It would be better to build in parallel from the very beginning, making many trillions of these molecules all at the same time. I think there is tremendous power in parallel assembly. Maybe another field, chemistry or biology, offers a better way to do it." The chemists at U. Lill aimed to develop that better way, building first simple and then more and more complex molecular machines. The eventual result was a primitive molecular assembler able to build molecular objects by the trillions.

Chemist's Tools

How did the chemists achieve this? During the years when the U. Brob team was developing the molecular manipulator, researchers working in protein science and synthetic chemistry had made better and better systems of molecular building blocks. Chemists were well prepared for doing this: by the late 1980s, it had become possible to build stable structures the size of medium-sized protein molecules, and work had begun to focus on making these molecules perform useful work by binding and modifying other molecules. Chemists learned to use these sophisticated catalysts-early molecular devices-to make their own work easier by helping in the manufacture of still more large molecules.

Another traditional chemist's tool was software for doing computer-aided design. The early software designed by *Jay Ponder* and *Frederic Richards* of Yale University ultimately led to semi-automatic tools for designing molecules of a particular shape and function. Chemists then could easily design molecules that would self-assemble into larger structures, several tens of nanometers across.

Molecular Construction Machines

These advances in software and chemical synthesis let the U. Lill team tackle the task of building a primitive version of a molecular assembler. Although they couldn't build anything as complex as a *nanocomputer* or as stiff as diamond, they didn't need to. Their design used sliding molecular rods to position a molecular gripper much like the gripper used at U. Brob, again using the surrounding liquid to control which tool the gripper held. Instead of an AFM's electronic controls, they used the surrounding liquid to control the position of the rods as well. In a neutral solution, the rods would withdraw; in an acid solution, they would extend. How far they moved depended on what other molecules were around to lodge in special pockets and block the motion.

Their primitive assemblers built much the same sorts of products that the U. Brob molecular manipulator did; the tools were similar, and speed and accuracy were about the same. Yet there was one dramatic advantage: About 1,000,000,000,000,000,000,000 U. Lill assemblers could fit in the space occupied by one U. Brob manipulator, and it was easy to produce a mere 1,000,000,000,000,000 times as much product at the same cost.

With the first, primitive assemblers, construction was slow because each step required new liquid baths and several seconds of soaking and waiting, and a typical product might take thousands of steps. Nonetheless, the U. Lill team made a lot of money licensing their technology to researchers trying to commercialize products they had

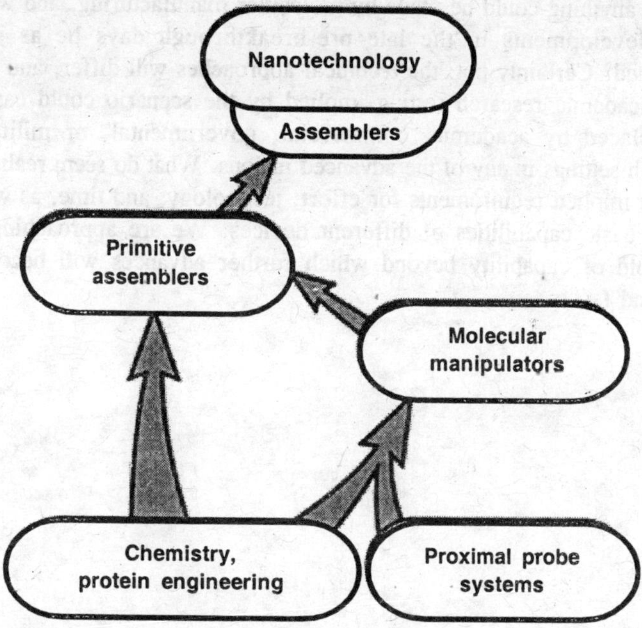

Fig. 7.2. Paths to nanotechnology.

first researched with the U. Brob machine. After starting an independent company (Nanofabricators), they poured their research efforts into building better machines.

Within a few years, they had assemblers with multiple grippers, each loaded with a different kind of tool; flashes of coloured light would flip molecules from state to state (they copied these molecules from the pigments of the retina of the eye); flipping molecules would change tools and change rod positions. Soaking and waiting become a thing of the past, and soon they were pouring out parts that, when mixed with liquid and added to dishes with special blank chips would build up the dense memory layers that made possible the Pocket Library. That was when things started moving fast. The semiconductor industry went the way of the vacuum tube industry. Money and talent poured into the new technology.

Molecular CAD tools got better, assemblers made it easy to build what was designed, and fast production and testing made molecular engineering as easy as playing with software. Assemblers got better, faster, and cheaper. Researchers used assemblers to build nanocomputers, and nanocomputers to control better, faster assemblers. Using tools to build better tools is an ancient story. Within a decade, almost anything could be made by molecular manufacturing, and was. Will developments in the late pre-breakthrough days be as just described? Certainly not: the technical approaches will differ, and the U.S. academic research setting implied by the scenario could easily be replaced by academic, commercial, governmental, or military research settings in any of the advanced nations. What do seem realistic are the implied requirements for effort, technology, and time, as well as the basic capabilities of different devices. We are approaching a threshold of capability beyond which further advances will become easy and fast.

8

Vision of Nanotechnology

The year is 2020. You are standing on a platform at the edge of a thousand-foot cliff overlooking the Tharsis volcano range on Mars. Your body casts a long shadow in the light of the setting sun as you scan the horizon for interesting features with your binoculars. Walking toward the edge of the precipice, you survey the texture of the rusty red boulders around you. Butterflies rise in your stomach as your peer over the cliff edge.

Sixteen hundred miles overhead, *Phobos*, one of the Martian moons, shines conspicuously. There is a new transport base under construction there. You move toward your telescope to get a better view when suddenly a doorway materializes out of empty space! A leg steps through it. It's your spouse with dinner.

Of course, you were never really on Mars. The "*Platform*" You were standing on was merely the floor of a small, comfortably furnished room called a *teleporter*. The walls and ceiling of this room are covered with one of the technological wonders of the twenty-first century: *phased array optics*.

Theory

Phased array optics is an application of nanotechnology that will produce three-dimensional views of objects and scenery using only two-dimensional displays. Display systems based on this technology—*optical phased arrays*—will behave quite literally as windows onto whatever scenery we can imagine.

Phased arrays are based on the theory of diffraction from physical optics. This theory says that patterns of light waves traveling beyond an aperture (such as a window) are entirely determined by the amplitude and phase distribution of light at the surface of the aperture. This means that if we produce light with the right phase and brightness distribution across a two-dimensional surface, we can reproduce the same light waves that would emanate from a three-dimensional scene behind the surface. In other words, we can make the surface appear as a window onto the scene.

Diffraction theory is mathematically continuous; it is assumed that surfaces can be reduced to an infinite number of small elements radiating with different amplitude and phase. This introduces complications form the standpoint of developing a practical technology. Fortunately we can achieve effective results with a discrete array of sources, provided they are coherent and less than half a wavelength apart. The wavelength of visible light ranges form 0 4 microns to 0.7 microns. A two-dimensional array of programmable sources 0.2 microns apart will therefore be sufficient to reconstruct any light wave pattern we desire.

Fig. 8.1 through 8.3 illustrate the reconstruction principle in two dimensions for a few simple cases. In Fig. 8.1 all the sources radiate

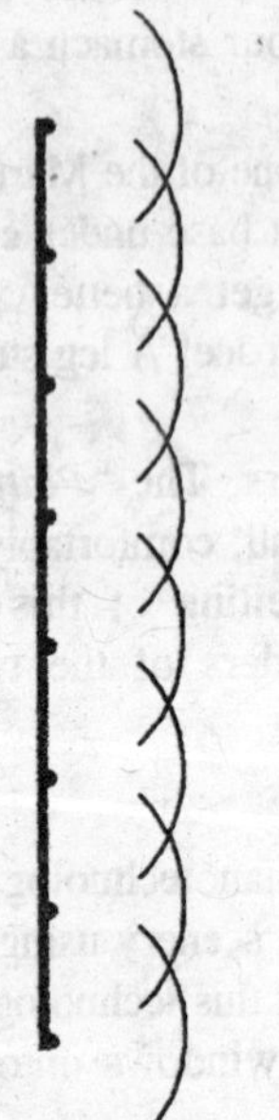

Fig. 8.1. Waves emanating from discrete sources in a phase array interface to create a plane wave that appears to come from a distant source behind the array.

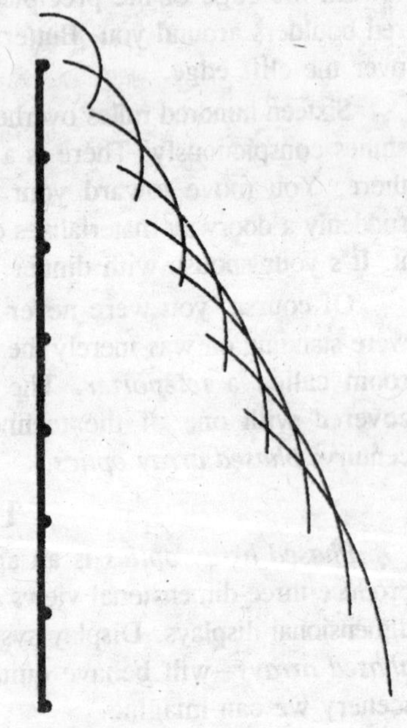

Fig. 8.2. By emitting waves with different phase relationships, the array sources are able to create a plane wave that propagates at any angle to the array surface.

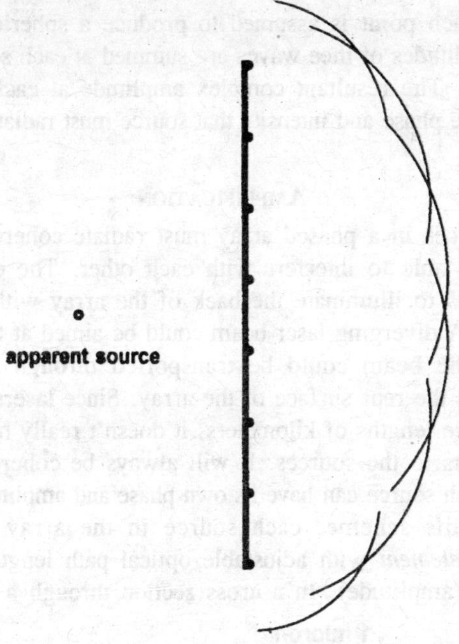

Fig. 8.3. A different combination of phase relationships produces a spherical wave that appears to come from a point near the array surface.

with the same amplitude and phase. Their waves interfere to create a plane wave propagating normal to the surface. This is the type of wave that would be produced by a distant point source. Secondly there is a linear variation of phase among the sources down the line. The result is a plane *wave propagating* at an angle to the surface. By adjusting the amount of phase variation, we can "*steer*" this beam in any direction we please. In Fig. 8.3 a more complicated phase relationship is used to produce a spherical wave, creating the image of a point source near the surface.

By choosing the correct phase and amplitude distribution across the array, we can, in fact, create images of any number of points at any number of locations behind the array. Since any three-dimensional scene can be represented as a collection of discrete points in space, it follows that an array can reproduce any three-dimensional scene.

The procedure for calculating the phase and amplitude distribution required to produce a scene is straightforward. All visible objects in the scene are represented in a computer in terms of discrete surface points. The spacing of the points is determined by the resolution of the array. Each point is assumed to produce a spherical wave. The complex amplitudes of thee waves are summed at each source position on the array. The resultant complex amplitude at each source point determines the phase and intensity that source must radiate to reproduce the scene.

Amplification

The sources in a phased array must radiate coherently. That is, they must be able to interfere with each other. The easiest way to achieve this is to illuminate the back of the array with light from a single laser. A diverging laser beam could be aimed at the array from behind, or the beam could be transported through a thin planar waveguide on the rear surface of the array. Since lasers can be made with coherence lengths of kilometers, it doesn't really matter how the laser light gets to the sources. It will always be coherent across the array, and each source can have its own phase and amplitude calibration factors. In this scheme, each source in the array is a *passive transmission element* with adjustable optical path length (phase) and transmission (amplitude). In a cross section through a few elements

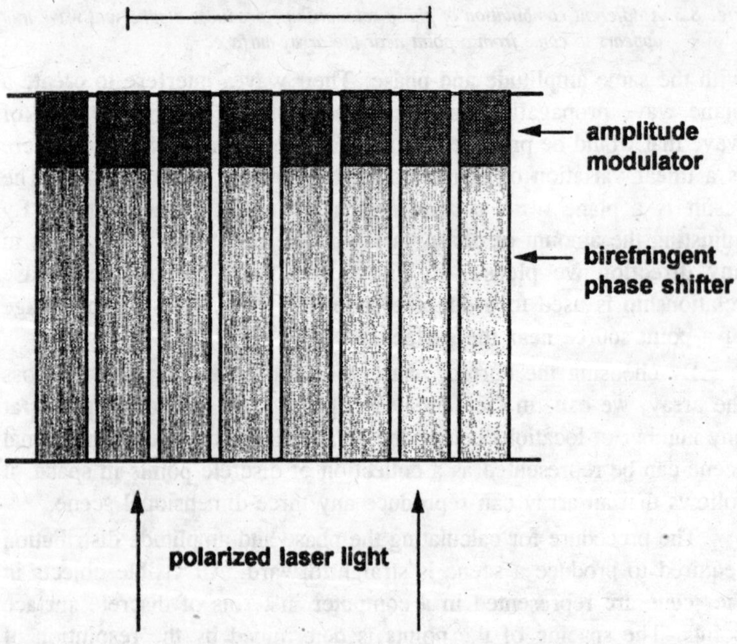

Fig. 8.4. Schematic cross section of an optical phased array.

seen, several elements placed 0.2 microns (200 nm) apart. (A bacterium could cover five of them.) Each element consists of a phase shifter and an amplitude modulator.

The amplitude modulator is simple. A cross polarizer like those in LCD displays should suffice. The phase *shifter* is made of an *electrooptic birefringent* material. This means that the shifter changes its refractive index along one polarization direction in response to an electric field. A polarizing filter down the back of the array ensures light of the proper linear polarization enters the shifter. A quarter-wave plate on the front of the array to restore elliptical polarization is optional.

A large variety of crystals are known to exhibit the required electrooptic effect. Some of these are semiconductors that are the stock in trade of existing microfabrication technology. Unfortunately the electrooptic effect in solid crystals is rather weal. Even very large electric fields produce very small changes in refractive index. Since each phase shifter must be able to retard light passing through it by one full wavelength, a small change in refractive index means a long phase shifter. For solid crystals this length will be on the order of a millimeter, which is probably too long to be practical.

Nematic liquid crystals exhibit an electrooptic effect thousands of times stronger than that found in solids. If these crystals can be made with a high refractive index and fast response, then phase shifters one micron long, should be possible. An interesting alternative has been suggested by Jeffrey Soreff of IBM. Rather than relying on electrooptic effects, a crystal with a large fixed birefringence, such as calcite, could be used. The crystal is sandwiched between parallel linear polarizers and rotated relative to the direction of polarization. The phase-shifting rotation could be accomplished by mechanical rotation of the crystal or by rotating the polarizer directions. This scheme would also permit micron-length phase shifters.

In all these designs each phase shifter is behaving as a waveguide near cutoff. *Red light*, with a wavelength in free space of 0.6 microns, will be the most difficult to deal with. To get through a waveguide less than 0.2 microns wide, the wavelength will have to be reduced to less than 0.4 microns. This will require a refractive index of at least 1.5 in the phase shifting medium.

Computational Management

Although the synthesis of array images may be mathematically simple, it will be computationally severe. A source separation of 0.2

microns gives 25 million sources per square millimeter, 2.5 billion per square centimeter, and 25 trillion array sources per meter. Each one of these sources will have to radiate with its own calculated phase and amplitude.

This is a staggering amount of information. To understand the origin of this information; recall that the array behaves as a window onto the scene it is reproducing. Suppose a square meter array is reproducing a wilderness scene. Looking at (through) the array, one sees a meadow, a small lake, and a mountain beyond. One mile away, on the far shore of the lake, a man is fishing. If you aimed a telescope in his direction, you would see him clearly. And if you had a really big telescope (one with a two-foot wide lens), you would be able to see a housefly on his hat. It is now obvious whey the array contains so much information. A meter wide array (or window for that matter) has a diffraction limited resolution of about one millionth of a radian, or 0.2 seconds of arc. The array image therefore contains every scene element subtending a solid angle greater than a trillionth of a steradian: the fly on the man's hat, leaves on trees five miles away, almost every blade of grass in the meadow, and so on.

The number of scene elements recorded by an array is roughly equal to the number of sources in the array. To compute the phase and amplitude for each source, we must add the contributions of spherical waves from each scene element to each source point. For a square meter array, this is 1000 trillion trillion floating point operations. A single computer would have trouble finishing this calculation within the lifetime of the universe.

Fortunately several considerations come to our rescue. First this type of calculation is amenable to Fast Fourier methods, which tend to reduce N-squared operations to NlogN operations (a trillion-fold decrease in our case). Second, this problem is well suited to massive parallelism. With enough processors, there is no reason why array image synthesis couldn't proceed at movie frame rates. Finally, not all array images will require resolution beyond that of the unaided eye. This concession alone reduces the computation and bandwidth requirements by several orders of magnitude.

Real-time arrays still pose formidable data storage and transmission problems. A square meter array operated at human eye resolution will require about a *gigabyte* per frame, or 50 gigabytes per second. Interestingly, light itself is ideally suited to transmission at this rate. Modulated light could be used for both long distance transmission and local control of array sources. Instead of trillions of tiny wires, light traveling through the same waveguide as the primary laser could carry image information to decoders at periodic interval in the array.

The *data storage* requirements of arrays cannot yet be met economically. (A phased array movie would consume several hundred terabytes!) Nevertheless, storage technologies able to handle array are on the horizon. Certainly molecular technology, with its promise of one gigabyte per cubic micron, will be more than sufficient.

Holography

Phased arrays and *holography* are both methods of wavefront reconstruction, and both can produce three-dimensional images. They differ in some important ways, however.

Holography avoids the computational requirement of arrays with a simple and elegant solution. The scene to be reproduced in illuminated with a laser. Photographic film is placed near the scene to capture reflected light (the "*object*" beam). At the same time, a reference beam from the same laser is shone on the film. Because they come from the same laser, the object and reference beams are coherent and produce a microscopic interference pattern that is recorded on the film.

This *interference pattern* is analogous to the phase and amplitude modulation of source in a phased array. When a laser beam is shown on the developed film, the transparent areas of the interference pattern select out those parts of the beam that have the correct phase to reconstruct the recorded scene.

Holography is simple and works without a computer; however, it has inherent drawbacks. Holographic recording produces phase modulation that is too coarse for unambiguous wave reconstruction. When you shine a laser on a *hologram*, three beams will always be emitted: the reconstructed object beam, the (unwanted) conjugate beam, and the laser beam itself. This happens because of the wide spacing of interference fringes produced by *holographic recording*. (The three beams are actually three orders of diffraction). *Phased arrays* modulate phase and amplitude at half-wavelength intervals. This is close enough to generate only one interference maximum, which contains the reconstructed beam and nothing else.

Simulated Incoherence

Phased arrays reproduce scenery using coherent light. This raises certain problems that have been glossed over so far. It was assumed earlier that different parts of a scene could be reproduced independently.

This is not quite true for coherent light. Light waves from different points in a coherent scene would interfere with each other, generating unnatural effects such as laser speckle.

A simple solution exists. Numerous separate versions of a scene could synthesized, each assigning a different random phase factor to points in the scene. When presented in rapid succession (say twenty versions per second), all interference effects would blur out, creating an effectively incoherent scene. In terms of eliminating laser speckle, this process is equivalent to rapidly altering the microscopic structure of a scene surface.

Colour

Producing coloured scenes is easy. Three versions of each scene would be synthesized in *red*, *green*, and *blue* light (the three primary colours of human vision). The array would cycle through them rapidly, alternating between red, green, and blue lasers. Because scenes would tend to split apart into separate colours if you moved your eyes fast, you would want a fast cycle time (ideally less than a millisecond).

Although they would appear perfectly authentic to human eyes, scenes produced in this way would be easily detectable by a prism. They might not fool animals either. The ultimate solution is to cycle very quickly through the entire visual spectrum at random increments. Such an array would be difficult to detect.

Transparent Planar

If a *transparent planar* waveguide is used to carry laser light across the back of a *phased array*, then the array will be semitransparent with perhaps 30 percent transmittance (like tinted glass). This raises all sorts of interesting possibilities. For one, the array could superpose images on real scenery behind it. (A glide path for a landing aircraft is a nice example). More profound, though, is what the array could do to the scenery itself.

In the theory of diffraction, almost every element of an optical system can be modeled as a two-dimensional surface with a complex transmission function. (A *lens*, for example, is equivalent to a flat surface with particular phase changing properties.) Phased arrays will turn this mathematical abstraction into technological reality. A single transparent array could behave as a programmable phase plate, Fresnel zone plate, hologram, prism, diffraction grating, or variable focal length, low of number, diffraction-limited lens. All this from an array microns thick and able to change roles in a microsecond. You could hold such a thing in your hand and swear it was from another world.

RECORDING PROBLEM

For many applications the scenery produced by an array will originate entirely in a computer. When it does not, a way must exist to record real scenery that allows reproduction by an array. In the general case of an incoherent (naturally illuminated) scene, the only way to do this is brute force: record conventional flat images from every angle you can get. For low-resolution reproduction, a "*fly's eye*" lattice of one-centimeter wide lenses could be used to record multiple views of a scene. A computer would then infer the three-dimensional structure of the scene from these views and proceed with image synthesis by methods already discussed. This system would have adequate resolution for viewing with the unaided eye. Higher resolution would require larger lenses. Transparent arrays are well suited to this application. A transparent array could behave as a fly's eye lens to record nearby objects and later become a single huge lens to record details miles away. This type of system could quickly gather all the information needed to reproduce scenes with *telescopic resolution*.

HIGH-POWER ARRAYS

The array designs presented so far operate by modulating laser light passing through them. The "*sources*" don't actually produce light themselves. For some applications, particularly outdoor use, a higher power design is desirable. The visual intensity of the sun (about 500 watts per square meter) is the regime we seek.

Consider putting a tiny *laser* at the base of each phase shifter. (Quantum well semiconductor lasers would be ideal). These lases provide a coherent boost to light passing through the array, essentially, transforming the whole array into a resonant cavity. In other words, the array becomes one big laser with adjustable emission phase across its surface. Colour could be dealt with by making the lasers tunable, or by laying them out in a red-green-blue hexagonal lattice.

Now consider a building covered on all sides by a high-power phased array. By day it could produce an image of a landscape contiguous with its surroundings, thereby rendering itself effectively invisible. By night it could do more of the same, or, for variety, create the image of a floodlit building. Perhaps it would supplement this scene with the image of a powerful, sweeping searchlight to detect and/or intimidate intruders. For those occasions when a searchlight is not intimidating enough, the phase of a few array sources could be matched to the coordinates of an intruder, with interesting consequences. Suppose an observer looking at the array sees one square meter worth of sources radiating at exactly the same phase relative to him. The sources will interfere constructively, depositing their entire output energy to a very small point at his location. In other words, the observer will be at the focus of a one-kilowatt laser (not a very healthy place to be).

Of course the same principle can be employed over the entire surface area of the building, encompassing hundreds of kilowatts of laser power. Our camouflage, decorative lighting system is now a long-range missile defense system, or directional transmitter of interstellar range. (With a beam divergence of less than a millionth of a degree, a stable 100-meter wide array would appear 1 percent as bright as the sun at any distance). While this example is a bit extreme, it illustrates that very small coherent sources can produce very big laser power. It follows that generating images of lasers and laser light will be an important application of phased arrays (particularly in view of the high efficiency of semiconductor lasers).

Passive and Active Invisibility

Displaying still scenery matching your surroundings is one form of invisibility. We could call this *passive invisibility*. With a view toward very fast and compact computers, an even more audacious possibility exists: active invisibility.

Active invisibility means an array would acquire and synthesize contiguous scenery in real time. This could be accomplished with system a few millimeters thick. The interior layer of our hypothetical "*invisibility suit*" consists of photosensors comparable to a human retina. The exterior is a phased array with phase shifters of extra close spacing and high refractive index. Half of the shifters are dedicated to image production, while the other half transmit light through to the photosensitive layer. The transmitting shifters are adjusted to form a multitude of focused flys's eye images on the photosensitive layer. The system is thus able to both produce and view scenery from all angles at the same time. A powerful computer network would manage the scenery synthesis and other details, such as how suit movement affected phase relationships. Other designs are possible. Such speculations lie at the limits of foreseeable technology. It is nonetheless amazing that something as fantastic as invisibility could be achieved using known science.

Front-Projection Images

Phased arrays can produce images of objects anywhere in space behind them. But that is not all they can do. With certain restrictions,

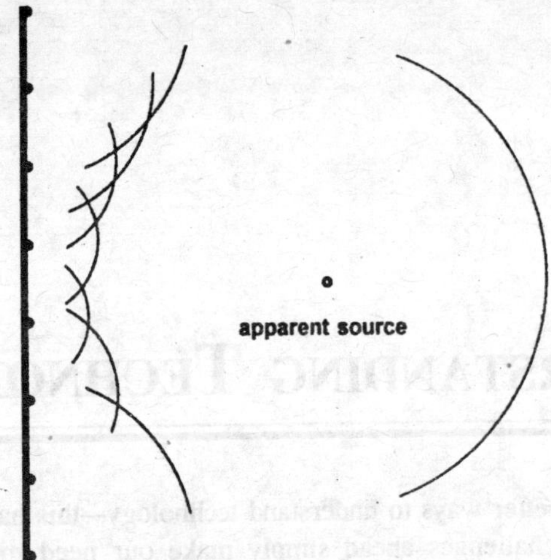

Fig. 8.5. Phased arrays can produce wavefronts that appear to originate in empty space in front to the array.

they can also make objects appear in front of them. By producing converging spherical waves, the array creates the image of a point source in front of it. Once again, creating a whole object is just a matter of assembling the points that constitute it. By controlling the areas of the array that reproduce various points, hidden lines and surfaces are removed, and even front-projected objects will appear solid and opaque. Imagine you are back in your teleporter room enjoying the Martian panorama. This time the floor is also covered with phased array optics. Martian boulders now appear at your feet, among the furniture, and in one case right in front of your nose! Now that's realism. *Virtual teleporters* covered on all sides by phased arrays bear a striking similarity to the "*holodeck*" concept of *Star Trek*. Anything could happen in such a room. Anything.

FUTURE

Optical phased arrays pose substantial challenges for information precessing and nanofabrication. Nevertheless it seems certain that these challenges will be met within the next twenty to thirty years. Some precursor technologies, such as *LCD holography*, are at the edge of current capability. Over the long term, the future of arrays is as limitless as imagination itself. The few applications mentioned here are just a glimpse of their potential. As Arthur C. Clarke said about virtual reality, this technology "won't merely replace TV. It will eat it alive."

9

UNDERSTANDING TECHNOLOGY

Society needs better ways to understand technology—this has long been obvious. The challenges ahead simply make our need more urgent. The promise of technology lures us onward, and the pressure of competition makes stopping virtually impossible. As the technology race quickens, new developments sweep toward us faster, and a fatal mistake grows more likely. We need to strike a better balance between our foresight and our rate of advance. We cannot do much to slow the growth of technology, but we can speed the growth of foresight. And with better foresight, we will have a better chance to steer the technology race in safe directions.

Various approaches to guiding technology have been suggested. "The people must control technology" is a plausible slogan, but it has two possible meanings. If it means that we must make technology serve human needs, then it makes good sense. But if it means that the people as a whole must make technical decisions, then it makes very little sense. The electorate cannot judge the intricate links between technology, economy, environment, and life; people lack the needed knowledge. The people themselves agree: according to a U.S. National Science Foundation survey, 85 percent of U.S. adults believe that most citizens lack the knowledge needed to choose which technologies to develop. The public generally leaves technical judgments to technical experts.

Unfortunately, leaving judgment to experts causes problems. In *Advice* and *Dissent*, Primack and von Hippel point out that "to the extent that the Administration can succeed in keeping unfavourable information quiet and the public confused, the public welfare can be sacrificed with impunity to bureaucratic convenience and private gain." Regulators suffer more criticism when a new drug causes a single death than they do when the *absence* of a new drug causes a thousand deaths. They misregulate accordingly. Military bureaucrats have a vested interest in spending money, hiding mistakes, and continuing their projects. They mismanage accordingly. This sort of problem is so basic and natural that more examples are hardly needed. Everywhere, secrecy and fog make bureaucrats more comfortable; everywhere, personal convenience warps factual statements on matters of public concern. As technologies grow more complex and important, this pattern grows more dangerous.

Some authors consider rule by secretive technocrats to be virtually inevitable. In *Creating Alternative Futures*, Hazel Henderson argues that complex technologies "become inherently totalitarian" (her italics) because neither voters nor legislators can understand them. In The *Human Future Revisited*, Harrison Brown likewise argues that the temptation to bypass democratic processes in solving complex crises brings the danger "that if industrial civilization survives it will become increasingly totalitarian in nature." If this were so, it would likely mean our doom: we cannot stop the technology race, and a world of totalitarian states based on advanced technology needing neither workers nor soldiers might well discard most of the population.

Fortunately, democracy and liberty have met comparable challenges before. States grew too complex for direct democracy, but representative government evolved. State power threatened to crush liberty, but the rule of law evolved. Technology has grown complex, but this gives us no reason to ignore the people, discard the law, and hail a dictator. We need ways to handle technical complexity in a democratic framework, using experts as instruments to clarify our vision without giving them control of our lives. But technical experts today are mired in a system of partisan feuding.

Establishment

Government and industry—and their critics—commonly appoint expert committees that meet in secret, if they meet at all. These committees claim credibility based on *who they are*, not on *how they work*. Opposed groups recruit opposed Nobel laureates.

To gain influence in our mass democracy, groups try to outshout one another. When their views have corporate appeal, they take them to the public through advertising campaigns. When their views have pork-barrel appeal, they take them to legislatures through lobbying.

When their views have dramatic appeal, they take them to the public through media campaigns. Groups promote their pet experts, the battle goes public, and quiet scientists and engineers are drowned in the clamor.

As the public conflict grows, people come to doubt expert pronouncements. They judge statements the obvious way, by their source. ("Of course she claims oil spills are harmless—she works for Exxon." "Of course he says Exxon lies—he works for Nader.")

When established experts lose credibility, demagogues can join the battle on an equal footing. Reporters—eager for controversy, striving for fairness, and seldom guided by technical backgrounds-carry all sides straight to the public. Cautious statements by scrupulous scientists make little impression; other scientists see no choice but to adopt the demagogues' style. Debates become sharp and angry, divisions grow, and the smoke of battle obscures the facts. Paralysis or folly often follows.

Our greatest problem is how we handle problems. Debates rage over the safety of nuclear power, coal power, and chemical wastes. Well-meaning groups backed by impressive experts clash again and again over dull, technical facts—dull, that is, save for their importance: What are the effects of low-level radiation, and how likely is a reactor meltdown? What are the causes and effects of acid rain? How well could space-based defenses block missile attacks? Do five cases of leukemia within three miles of a particular waste dump show a deadly hazard, or merely the workings of chance?

Greater issues lie ahead: How safe is this replicator? Will this active shield system be safe and secure? Will this biostasis procedure be reversible? Can we trust this AI system?

Disputes about technical facts feed broader disputes about policy. People may have differing values (which would you rather have, encephalitis or pesticide poisoning?) but their views of relevant facts often differ still more. (How often do these mosquitoes carry encephalitis? How toxic is this pesticide?) When different views of boring facts lead to disagreements about important policies, people may wonder, "How can they oppose us on this vital issue unless they have bad motives?" Disputes over facts can thus turn potential allies against one another. This hampers our efforts to understand and solve our problems.

People have disputed facts for millennia; only the prominence of technical disputes is new. Societies have evolved methods for judging facts about people. These methods suggest how we might judge facts about technology.

Trial and Hit

Throughout history, groups have evolved ways to resolve disputes; the alternative has been feuds, open-ended and often deadly. Medieval Europeans used several procedures, all better than endless feuding: They used trial by battle: opponents fought, and the law vindicated the victor.

They used compurgation: neighbours swore to the honesty of the accused; if enough swore, the charges were dropped. They used trial by ordeal: in one, the accused was bound and thrown in a river; those who sank were innocent, those who floated, guilty.

They used judgment by secretive committees: the king's councilors would meet to judge and pass sentence as seemed fit. In England, they met in a room called the Star Chamber.

These methods supposedly determined *who did what*—the facts about human events—but all had serious shortcomings. Today we use similar methods to determine *what causes what*—the facts about science and technology:

We use trial by combat in the press: opponents fling sharp words until one side's case suffers political death. Unfortunately, this often resembles an endless feud.

We use compurgation: experts swear to certain facts; if enough swear the same, the facts are declared true. We use judgment by secretive committees: selected experts meet to judge facts and recommend such actions as seem fit. In the United States, they often meet in committees of the National Academy of Sciences.

Trial by ordeal has passed from fashion, but combat in the press may well seem like torture to the quiet scientist with self-respect.

The English abolished Star Chamber proceedings in 1641, and they counted this a great achievement. Trial by combat, compurgation, and ordeal have likewise become history. We now value due process, at least when judging people.

Court procedures illustrate the principles of due process: Allegations must be specific. Both sides must have a chance to speak and confront each other, to rebut and cross-examine. The process must be public, to prevent hidden rot. Debate must proceed before a jury that both sides accept as impartial. Finally, a judge must referee the process and enforce the rules. To see the value of due process, imagine its opposite: a process trampling all these principles would give one side a say and the other no chance to cross-examine or respond. It would meet in secret, allow vague smears, and lack a judge to enforce whatever rules might remain. Jurors would arrive with their decisions made. In short, it would resemble a lynch mob meeting in a locked barn—or a rigged committee drafting a report.

Experience shows the value of due process in judging facts about people; might it also be of value in judging facts about science and technology? Due process is a basic idea, not restricted to courts of law. Some AI researchers, for example, are building due-process principles into their computer programs. It seems that due process should be of use in judging technical facts. In fact, the scientific literature—the chief forum of science—already embodies a form of due process: In good journals, scientific statements must be specific. In theory, given enough time and persistence, all sides may state their views in a dispute, since journals stand open to controversy. Though opponents may not meet face to face, they confront each other at a distance; they question and respond in slow motion, through letters and articles. Referees, like juries, evaluate evidence and reasoning. Editors, like judges, enforce rules of procedure. Publication keeps the debate open to public scrutiny.

In both journals and courts, conflicting ideas are pitted against one another under rules evolved to ensure a fair, orderly battle. These rules sometimes fail because they are broken or inadequate, but they are the best we have developed. Imperfect due process has proved better than no due process at all.

Why do scientists value refereed journals? Not because they trust all refereed journals, or trust everything printed in any one of them. Even the best due-process system won't grind out a stream of pure truth. Rather, they value refereed journals because they tend to reflect sound critical discussion. Indeed, they must: because journals compete with one another for papers, prestige, and readership, the best journals must be good indeed. Journals grind slowly, yet after enough rounds of publication and criticism they often grind out consensus. Experience proves the value of both courts and journals. Their underlying similarity suggests that their value stems from a common source—due process. Due process can fail, but it is still the best approach known for finding the facts.

Today, courts and journals are not enough. Vital technical disputes go on and on because we have no rapid, orderly way to bring out the facts (and to delineate our ignorance). Courts are not suited to deal with technical questions. Journals are better, but they still have shortcomings. They took shape in a time of lower technology and slower advance, evolving to fit the limits of printing, the speed of mail, and the needs of academic science. But today, in a time when we desperately need better and swifter technical judgment, we find ourselves in a world that has telephones, jets, copiers, and express and electronic mail. Can we use modern technologies to speed technical debate?

Of course: scientists already use several approaches. Jets bring scientists from around the globe to conferences where papers are presented and discussed. But conferences handle controversy poorly; public decorum and tight schedules limit the vigor and depth of debate. Scientists also join informal research networks linked by telephone, mail, computers, and copying machines; these accelerate exchange and discussion. But they are essentially private institutions. They, too, fail to provide a credible, public procedure for thrashing out differences.

Conferences, journals, and informal networks share some similar limitations. They typically focus on technical questions of scientific importance, rather than on technical questions of public-policy importance. Moreover, they typically focus on *scientific* questions. Journals tend to slight technological questions that lack intrinsic scientific interest; they often treat them as news items not worthy of checking by referees. Further, our present institutions lack any balanced way to present knowledge when it is still tangled in controversy. Though scientific review articles often present and weigh several sides, they do so from a single author's point of view.

All these shortcomings share a common source: scientific institutions evolved to advance science, not to sift facts for policy makers. These institutions serve their purpose well enough, but they serve other purposes poorly. Though this is no real failing, it does leave a real need.

Better Procedure

We need better procedures for debating technical facts—procedures that are open, credible, and focused on finding the facts we need to formulate sound policies. We can begin by copying aspects of other due-process procedures; we then can modify and refine them in light of experience. Using modern communications and transportation, we can develop a focused, streamlined, journal-like process to speed public debate on crucial facts; this seems half the job. The other half requires distilling the results of the debate into a balanced picture of our state of knowledge (and by the same token, of our state of ignorance). Here, procedures somewhat like those of courts seem useful.

Since the procedure (a *fact forum*) is intended to summarize facts, each side will begin by stating what it sees as the key facts and listing them in order of importance. Discussion will begin with the statements that head each side's list. Through rounds of argument, cross examination, and negotiation the referee will seek agreed-upon statements. Where disagreements remain, a *technical panel* will then write opinions, outlining what seems to be known and what still seems uncertain. The output of the fact forum will include background arguments, statements of agreement, and the panel's opinions. It might resemble a set of journal articles capped by a concise review article - one limited to factual statements, free of recommendations for policy. This procedure must differ from that of a court in various ways. For example, the technical panel—the forum's "jury"—must be technically competent. Bias might lead a panel to misjudge facts, but technical incompetence would do equal harm. For this reason, the "jury" of a fact forum must be selected in a way that might be dangerous if allowed in courts of law. Since courts wield the power of the police, we use juries selected from the people as a whole to guard our liberty.

This forces the government to seek approval from a group of citizens before it punishes someone, thus tying the government's actions to community standards. A fact forum, however, will neither punish people nor make public policy. The public will be free to watch the process and decide whether to believe its results. This will give people control enough.

Still, to make a fact forum fair and effective, we will need a good panel-selection procedure. Technical panels will correspond roughly to the expert committees appointed by governments or to the referees appointed by journals. To ensure fairness, a panel must be selected not by a committee, a politician, or a bureaucrat, but by a process that involves the consent of both sides in the dispute. In court proceedings, advocates can challenge and reject any jurors who seem biased; we can use a similar process in selecting the panel for a fact forum. Experts who are directly involved in a dispute can't serve on the panel - they would either bias the panel or split it. The group sponsoring a fact forum must seek panelists who are knowledgeable in related fields. This seems practical because the methods of technical judgment (often based on experiments and calculations) are quite general. Panelists familiar with the fundamentals of a field will be able to judge the detailed arguments made by each side's specialists.

Other parts of the fact forum will also resemble those of courts and journals. A committee like a journal's editorial group will nominate a referee and panelists for a dispute. Advocates for each side, like authors or attorneys, will assemble and present the strongest case they can.

Despite these similarities, a fact forum will differ from a court: It will focus on technical questions. It will suggest no actions. It will lack government power. It will follow technical rules of evidence and argument. It will differ in endless details of tone and procedure. The analogy with a court is just that - an analogy, a source for ideas.

A fact forum will also differ from a journal: It will move as fast as mail, meetings, and electronic messages permit, rather than delaying exchanges by many months, as in typical journal publication. It will be convened around an issue, rather than being established to cover a scientific field. It will summarize knowledge to aid decisions, rather than serving as a primary source of data for the scientific community. Although a series of fact forums won't replace a journal, they will help us find and publicize facts that could save our lives. Dr. Arthur Kantrowitz (a member of the National Academy of Sciences who is accomplished in fields ranging from medical technology to high-power lasers) originated the concept I have just outlined. He at first called it a "board of technical inquiry." Journalists promptly dubbed it a "science court." I have called it a "fact forum"; I will reserve the term "science court" for a fact forum used (or proposed) as a government institution. Proposals for due process in technical disputes are still in flux; different discussions use different terms.

Dr. Kantrowitz's concern with due process arose out of the U.S. decision to build giant rockets to reach the Moon in one great leap; he, backed by the findings of an expert committee, had recommended that NASA use several smaller rockets to carry components into a low orbit, then plug them together to build a vehicle to reach the Moon. This approach promised to save billions of dollars and develop useful space-construction capabilities as well. No one answered his arguments, yet he failed to win his case. Minds were set, politicians were committed, the report was locked in a White House safe, and the debate was closed. The technical facts were quietly suppressed in the interests of those who wanted to build a new generation of giant rockets.

This showed a grave flaw in our institutions—one that persists, wasting our money and increasing the risk of a disastrous error. Dr. Kantrowitz soon reached the now obvious conclusion: we need due process institutions for airing technical controversies.

Dr. Kantrowitz pursued this goal (in its science-court form) through discussions, writings, studies, and conferences. He won endorsements for the science court idea from Ford, Carter, and Reagan—as candidates. As Presidents, they did nothing, though a presidential advisory task force during the Ford administration did detail a proposed procedure.

Still, progress has been made. Although I have used the future tense in describing the fact forum, experiments have begun. But before describing a path to due process, it makes sense to consider some of the objections.

Facts and Critics

Critics of this idea (at least in its science court version) have often disagreed with one another. Some have objected that factual disputes are unimportant, or that they can be smoothed over behind closed doors; others have objected that factual disputes are too deep and important for due process to help. Some have warmed that science courts would be dangerous; others have warned that they would be impotent. These criticisms all have some validity: due process will be no cure-all. Sometimes it won't be needed, and sometimes it will be abused. Still, one might equally well reject penicillin on the grounds that it is sometimes ineffective, unnecessary, or harmful.

These critics propose no alternatives, and they seldom argue that we have due process today, or that due process is worthless. We must deal with complex, technical issues on which millions of lives depend; dare we leave these issues to secretive committees, sluggish journals, media battles, and the technical judgment of politicians? If we distrust experts, should we accept the judgment of secretive committees appointed in secret, or demand a more open process? Finally, can we with our present system cope with a global technology race in nanotechnology and artificial intelligence?

Open, due-process institutions seem vital. By letting all sides participate, they will harness the energy of conflict to a search for the facts. By limiting experts to describing the facts, they will help us cope with technology without surrendering our decisions to technocrats. Individuals, companies, and elected officials will keep full control of policy; technical experts will still be able to recommend policies through other channels.

How can we distinguish facts from values? Karl Popper's standard seems useful: a statement is factual (whether true or false) if an experiment or observation could in principle disprove it. To some people, the idea of *examining facts* without considering values suggests the idea of *making policy* without considering values. This would be absurd: by their very nature, policy decisions will always involve both facts and values. Cause and effect are matters of fact, telling us what is possible. But policy also involves our values, our motives for action. Without accurate facts, we won't get the results we seek, but without values - without desires and preferences—we wouldn't seek anything in the first place. A process that uncovers facts can help people choose policies that will serve their values.

Critics have worried that a science court will (in effect) declare the Earth to be flat, and then ignore an Aristarchus of Samos or a Magellan when he finds evidence to the contrary. Errors, bias, and imperfect knowledge will surely cause some memorable mistakes. But members of a technical panel need not claim a bogus certainty. They can instead describe our knowledge and outline our ignorance, sometimes stating that we simply don't know, or that present evidence gives only a rough idea of the facts. This way, they will protect their reputations for good judgment. When new evidence arrives, a question can he reopened; ideas need no protection from double jeopardy.

If fact forums become popular and respected, they will gain influence. Their success will then make them harder to abuse: many competing groups will sponsor them, and a group that abuses the procedure will tend to gain a bad reputation and be ignored. No single sponsoring group will be able to obscure the facts about an important issue, if fact forums gain reliability through redundancy.

No institution will be able to eliminate corruption and error, but fact forums will be guided, however imperfectly, by an improved standard for the conduct of public debate; they will have to fall a long way before becoming worse than the system we have today. The basic case for fact forums is that (1) due process is the right approach to try, and (2) we will do better if we try than if we don't.

Public Benefits

Anthropologist Margaret Mead was invited to a colloquium on the science court to speak against the idea. But when the time came, she spoke in its favour, remarking that "We need a new institution. There isn't any doubt about that. The institutions we have are totally unsatisfactory. In many cases, they are not only unsatisfactory, they involve a prostitution of science and a prostitution of the decision making process." People with no vested interest in the existing institutions often agree with her evaluation.

If finding the facts about technology really is crucial to our survival, and if due process really is the key to finding the facts, then what can we do about it? We needn't begin with perfect procedures; we can begin with informal attempts to improve on the procedures we have. We can then evolve better procedures by varying our methods and selecting those that work best. Due process is a matter of degree.

Existing institutions could move toward due process by modifying some of their rules and traditions. For example, government agencies could regularly consult opposing sides before appointing the members of an expert committee. They could guarantee each side the right to present evidence, examine evidence, and cross-examine experts. They could open their proceedings to observers. Each of these steps would strengthen due process, changing Star Chamber proceedings into institutions more worthy of respect.

The public benefits of due process won't necessarily make it popular among the groups being asked to change, however. We haven't heard the thunder of interest groups rushing to test their claims, nor the cries of joy from committees as they throw open their doors and submit to the discipline of due process. Nor have we heard reports of politicians renouncing the use of spurious facts to hide the political basis of their decisions.

Yet three U.S. presidential candidates did endorse science courts. The Committee of Scientific Society Presidents, which includes twenty-eight of the leading scientific societies in the U.S., also endorsed the idea. The U.S. Department of Energy used a "science court-like procedure" to evaluate competing fusion-power proposals, and declared it efficient and useful. Dr. John C. Bailar of the National Cancer Institute, after failing to make medical organizations recognize the dangers of X rays and reduce their use in mass screening, proposed holding a science court on the subject. His opponents then backed down and changed their policies—apparently, the mere threat of due process is already saving lives. Nevertheless, the old ways continue almost unchallenged. Why is this? In part, because knowledge is power, and hence jealously guarded. In part, because powerful groups can readily imagine how due process would inconvenience them. In part, because an effort to improve problemsolving methods lacks the drama of a campaign to fight problems directly; a thousand activists bail out the ship of state for every one who tries to plug the holes in its hull.

Governments may yet act to establish science courts, and any steps they may take toward due process merit support. Yet it is reasonable to fear government sponsorship of science courts: centralized power tends to beget lumbering monsters. A central "Science Court Agency" might work well, do little visible harm, and yet impose a great hidden cost: its very existence (and the lack of competition) might block evolutionary improvements. Other paths lie open. Fact forums will he able to exert influence without help from special legal powers. To have a powerful effect, their results only need to be more credible than the assertions of any particular person, committee, corporation, or interest group. A well-run fact forum will carry credibility in its very structure. Such forums could be sponsored by universities. In fact, Dr. Kantrowitz recently conducted an experimental procedure at the University of California at Berkeley. It centered around public disputes between geneticist Beverly Paigen and biochemist William Havender regarding birth defects and genetic hazards at the Love Canal chemical dump site. They served as advocates; graduate students served as a technical panel. The procedure involved meetings spread over several weeks, chiefly spent in discussing areas of agreement and disagreement; it wound up with several sessions of public cross-examination before the panel. Both the advocates and the panel agreed to eleven statements of fact, and they clarified their remaining disagreements and uncertainties.

Arthur Kantrowitz and Roger Masters note that "in contrast to the difficulties experienced in the many attempts to implement a science court under government auspices, encouraging results were obtained in the first serious attempt... in a university setting." They remark that the traditions and resources of universities make them natural settings for such efforts. They plan more such experiments.

This shows a decentralized way to develop due-process institutions, one that will let us outflank existing bureaucracies and entrenched interests. In doing this, we can build on established principles and test variations in the best evolutionary tradition.

Leaders concerned with a thousand different issues—even leaders on opposing sides of an issue—can join in support of due process. In Getting to Yes, Roger Fisher and William Ury of the Harvard Negotiation Project point out that opponents tend to favour impartial arbitration when each side believes itself right. Both sides expect to win, hence both agree to participate. Fact forums should attract honest advocates and repel charlatans. As due process becomes standard, advocates with sound cases will gain an advantage even if their opponents refuse to cooperate—"If they won't defend their case in public, why listen to them?" Further, many disputes will be resolved (or avoided) without the trouble of actual proceedings: the prospect of a refereed public debate will encourage advocates to check their facts before they take stands. Establishing fact forums could easily do more for a sound cause than would the recruitment of a million supporters.

Yet today well-meaning leaders may feel forced to exaggerate their cases simply to be heard above the roar of their opponents' press releases. Should they regretfully oppose the discipline of fact forums? Surely not; due process can heal the social illness that forced them away from the facts in the first place. By making their points in a fact forum, they can recover the selfrespect that comes with intellectual honesty. Unearthing the truth may undermine a cherished position, but this cannot harm the interests of a true publicinterest group. If one must move a bit to make room for the truth, so what? Great leaders have shifted positions for worse reasons, and due process will make opponents shift too.

Gregory Bateson once stated that "no organism can afford to be conscious of matters with which it could deal at unconscious levels." In the organism of a democracy, the conscious level roughly corresponds to debate in the mass media. The unconscious levels consist of whatever processes ordinarily work well enough without a public hue and cry. In the media, as in human consciousness, one concern tends to drive out another. This is what makes conscious attention so scarce and precious. Our society needs to identify the facts of its situation more swiftly and reliably, with fewer distracting feuds in the media. This will free public debate for its proper task—judging procedures for finding facts, deciding what we want, and helping us choose a path toward a world worth living in.

As change quickens, our need grows. To judge the risk of a replicator or the feasibility of an active shield, we will need better ways to debate the facts. Purely social inventions like the fact forum will help, but new social technologies based on computers also hold promise. They, too, will extend due process.

10

Safety, Accidents, and Abuse

Some truisms: Almost any technology is subject to use, misuse, abuse, and accidents. The more powerful a technology is when properly used, the worse it is likely to be when abused. Any powerful technology in human hands can be the subject of accidents. *Nanotechnology* and *molecular manufacturing* will be no exception. Indeed, if molecular manufacturing replaces modern industry, and if its nanotechnological products replace most modern technologies, then most future accidents will have to involve nanotechnology.

Another truism: In a diverse, competitive world, any reasonably-inexpensive technology with enormous commercial, medical, and military applications will almost surely be developed and used. It is hard to envision a scenario (short of the collapse of civilization) in which nanotechnology will not make its appearance; it seems inevitable. If so, then its problems, however tough, must be dealt with.

Like trucks, aircraft, biotechnology, rockets, computers, boots, and warm clothes, nanotechnology has the potential for both peaceful and aggressive uses. In peaceful uses (by definition), harm to people occurs either by accident or as an unintended consequence. In aggressive uses, harm is deliberate. In a peaceful context, the proper question to ask is *Can fallible people of goodwill, pursuing normal human purposes, use nanotechnology in a way that reduces risk and harm to others?* In an aggressive, military context, the proper question to ask is *Can we somehow keep the peace?* Our answer to the first will be a clear yes, and to the second, an apprehensive maybe.

Throughout this discussion, we assume that most people will be alert in matters concerning their personal safety, and that some will be alert in matters concerning world safety. During the 1970s, people awakening to the new large-scale, long-term problems of technology often felt isolated and powerless. They naturally felt that technology was out of their control, in the hands of shortsighted and irresponsible groups. Today, there are still battles to be fought, but the tide has turned. When a concern arises regarding a new, obvious technology, it is now much easier to get a hearing in the media, in the courts, and in the political arena. Improving these mechanisms for social vigilance and the political control of technology is an important challenge. Current mechanisms are imperfect, but they can still give a big push in the right directions.

Though we assume alertness, alertness can be a scarce resource. The total amount of concern and energy available for focusing on long-term problems is so limited that it must be used carefully, not squandered on problems that are trivial or illusory. Part of our aim in this chapter is to help sort out the issues raised by nanotechnology so that attention can be focused on problems that must be solved, but might not be.

The next few sections deal with accidents of conventional sorts, where safety benefits are obvious. Later sections discuss more novel problems, some tough enough that we have no good answers.

Safety in Ordinary Activities

As countries have grown richer, their people have lived longer despite pollution and automobile accidents. Greater wealth means safer roads, safer cars, safer homes, and safer workplaces. Throughout history, new technologies have brought new risks, including risks of death, injury, and harm to the environment, but prudent people have only accepted new technologies when they offered an improved mix of risks and benefits. Despite occasional dramatic mistakes, the historical record says that people have succeeded in choosing technologies that reduce their personal risks. This must be so, or we wouldn't be living longer.

Molecular manufacturing and its products should continue this trend, not as an automatic consequence, but as a result of continued vigilance, of people exercising care in picking and choosing which technologies they allow into their daily lives. Nanotechnology will give better control of production and products, and better control usually means greater safety. Nanotechnology will increase wealth, and safety is a form of wealth that people value. Public debate, product testing, and safety regulations are standard parts of this process.

Home Safety

In common home accidents, a dangerous product is wrongly applied, spilled, or consumed. Homes today are full of corrosive and toxic materials, for cleaning drains, dissolving stains, poisoning insects, and so forth. All too often, children drink them and die. With advanced technology, none of these tasks will require such harsh, crude chemicals. Cleaning could be performed by selective *nanomachines* instead of corrosive chemicals; insects could be controlled by devices like *ecosystem protectors* that know the difference between a cockroach and a person or a ladybug. There will doubtless be room for deadly accidents, but with care and hard work, it should be possible to ensure that nanotechnologies for the home are safer than what they replace, saving many lives.

It is, of course, possible to imagine safety nightmares: nanotechnology could be used to make products far more destructive than anything we've seen because it could be used to extend almost any ability further than we've seen. Such products presumably won't be commonplace: even today, nerve gas would make a potent pesticide, but it isn't sold for home use. Thinking realistically about hazards requires common sense.

Industrial Safety

We've already seen how post-breakthrough technologies can eliminate oil spills by eliminating oil consumption. A similar story could be told of almost any class of industrial accident today. But what about accidents—spills and the like—with the new technologies? Rather than trying to paint a picture of a future technology, of how it could fail and what the responses could be, it seems better to try a thought experiment. What could be done to deal with oil spills, if oil were still in use? This will show how nanotechnologies can be used to cope with accidents:

If there were a spill and oil on the shore, advanced nanomechanisms could do an excellent job of separating oil from sand, removing oil from rocks, and cleaning crude oil from feathers on birds and the feathery legs of barnacles. Oil contamination is a pollution problem, and nanotechnology will be a great aid in cleaning up pollution.

But why should the oil reach the shore? Economical production would make it easy to stockpile cleanup equipment near all the major shipping routes, along with fleets of helicopters to deliver it at the first distress call from a tanker. Oil cleanup equipment built with nanotechnology could surely do an excellent job of scooping oil from the water before it could reach the shore.

But why should the oil leave the tanker? Economical production of strong materials could make seamless hulls of fibrous materials far tougher than steel, with double, triple, or quadruple layers. *Smart materials* could even make punctures self-sealing. Hulls like this could be run into rocks at highway speeds without spilling oil.

But why should anyone be shipping crude oil across the sea? Even if oil were still being pumped (despite inexpensive solar energy and solar-derived fuels), efficient *molecule* processing systems could refine it into pure, fluid fuels at the wellhead, and inexpensive tunneling machines could provide routes for deeply buried pipelines.

Any one of these advances would shrink or eliminate today's problem with oil spills, and all of them are feasible. This example suggests a general pattern. If nanotechnology can provide so many different ways to avoid or deal with an oil spill—one of the largest and most environmentally destructive accidents caused by today's industry—it can probably do likewise for industrial accidents in general.

The most direct approach is the most basic: the elimination of anything resembling today's bulk industrial plants and processes. The shift from messy drilling activities and huge tankers to small-scale distributed systems based on solar *cells* is characteristic of the style in which nanotechnology can be used. The chemical industry today typically relies on plants full of large, pressurized tanks of chemicals. Not surprisingly, these occasionally spill, explode, or burn. With nanotechnology, chemical plants will be unnecessary because molecules can be transformed in smaller numbers, as needed and where needed, with no need for high temperatures, high pressures, or big tanks. This will not only avoid polluting by-products, but reduce the risk of accidents.

Medical Safety

Medicine can be safer too. Drugs often have side effects that can do permanent damage or kill. Nanomedicine will offer alternatives to flooding the body with a possibly toxic chemical. Often, one wants to affect just one target: just the stomach, or perhaps just the ulcer. An antibiotic or antiviral treatment should fight specific *bacteria* or *viruses* and not damage anything else. When medicine achieves the sophistication of *immune machines* and *cell-surgery* devices, this will become possible.

But what about medical accidents and side effects? Molecular manufacturing will make possible superior sensors to tell medical researchers of the effects of a new treatment, thereby improving testing. Better sensors will also help in monitoring any negative effects of a treatment on an individual patient. With care, only a few cells would be damaged and only small concentrations of toxic by-products would be produced before this was noticed and the treatment corrected.

The resources of nanotechnology-based medicine would then be available for dealing with the problem. With biostasis techniques available, even the worst medically induced illnesses could be put on hold while a treatment was developed. In short, serious medical mistakes could be made far rarer, and most mistakes could be corrected.

The conclusion that follows from these examples of oil spills, chemical plants, and the effects of medical treatments is straightforward. Today our comparative poverty and our comparative technological incompetence press us in the direction of building and using relatively dangerous and destructive devices, systems, and techniques. With greater wealth and technological competence, we will have the option of accomplishing what we do today (and more) with less risk and less environmental destruction: in short, being able to do more, and do it better.

With better-controlled technologies, and with an ample measure of foresight and concern, we will even be able to do a better job of recovering from mistakes. It won't happen automatically, but with normal care we can arrange for our future accidents to be smaller and less frequent than those in our past.

Extraordinary Accidents

The previous section discussed ordinary accidents that would occur during the use of nanotechnology by generally responsible, yet fallible, human beings. Nanotechnology also raises the specter, however, of what have been termed "extraordinary accidents": accidents involving runaway self-replicating machines. One can imagine building a device about the size of a bacterium but tougher and more nearly omnivorous. Such runaways might blow like pollen and reproduce like bacteria, eating any of a wide range of organic materials: an ecological disaster of unprecedented magnitude—indeed, one that could destroy the biosphere as we know it. This may be worth worrying about, but can this happen by accident?

How to Prepare a Big Mistake

The so-called "*Star Trek scenario*" (named after an episode of *Star Trek: The Next Generation* that featured runaway "nanites") is perhaps the most commonly imagined problem. In this scenario, someone first invests considerable engineering effort in designing and building devices almost exactly like the one just described: bacterial-sized, omnivorous, able to survive in a wide range of natural environments, able to build copies of themselves, and made with just a few built-in safeguards—perhaps a clock that shuts them off after a time, perhaps something else. Then, accidentally, the clock fails, or one of these dangerous *replicators* builds a copy with a defective clock, and away we go with an unprecedented ecological disaster.

This would be an extraordinary accident indeed. Note well, though, that this accident scenario starts with someone building a highly capable device that is almost disastrously dangerous, but held in check by a few safeguards. This would be like wiring your house with dynamite and relying on a safety-catch to protect the trigger: a subsequent explosion could be called an accident, but the problem isn't with the safety mechanism, it's with the dynamite installation.

Do we need to build nanotechnological dynamite? It's worth considering just how little practical incentive there is for anything even resembling the dangerous replicator just described. (Note that our topic here is accidents; deliberate acts of aggression are another matter.)

How to Avoid It

With our present technology, which is simpler to build—a car that runs on gasoline, or one that forages for fuels in the forest? A foraging car would be very hard to design, cost more to manufacture, and have more parts to break down. The situation is similar with nanotechnology.

Ralph Merkle of Xerox Palo Alto Research Center discussed this issue at the *First Foresight Conference on Nanotechnology*. He explains, "It's both uneconomical and more difficult to design a self-replicating system that manufactures every part it needs from naturally occurring compounds. Bacteria do this, but in the process they have to synthesize all twenty amino acids and many other compounds, using elaborate enzyme systems tailored specifically for the purpose. For bacteria facing a hostile world, the ability to adapt and respond to a changing environment is worth almost any cost, for lacking this ability they would be wiped out.

"But in a factory setting, where adequate supplies of all the needed parts are provided, the ability to synthesize parts from scratch is not only unneeded, it consumes extra time and energy, and produces excess waste. Even if we could design artificial self-replicating systems as flexible as existing natural ones, an inflexible and rigid system is better adapted to the controlled factory setting in which it will find itself than a more complex, more adaptable, less efficient design."

What is more, the Desert Rose Industries scenario showed how an expandable factory setup could operate with no self-replicating machines at all: molecular manufacturing doesn't require them. If they are used for some purpose, they will most likely resemble automobiles in their finicky requirements. A self-replicating *molecular machine* built for industrial purposes (and made as simple as possible) would float in a container of specially selected chemicals. As with the automobile, the best chemicals to use will probably be chemicals not commonly found in nature, and it would be easy to make that a design rule: *Never make a replicator that can use an abundant natural compound as fuel.*

If we follow this rule, the idea of a replicator "escaping" and replicating in the wild will be as absurd as the notion of an automobile going feral and refueling itself from tree sap. Whether for replicators or cars, to design a machine that could operate in the wild would not be a matter of a flick of the draftsman's pen, but an intense, sustained research-and-development effort focused on that objective. Crashes and explosions occur in machinery by accident, but complex new capabilities don't.

A simple psychological error frequently occurs when someone first hears about nanotechnology, and hears mention of "molecular machines," and "replicators," and "*nanocomputers*," and "nanomachines that operate in nature." The error is this: The person makes a single new mental pigeonhole for "nanotechnology," throws everything into it, and stirs. After some mental fermentation, the result is the mythical nanomachine that does everything: it's a replicator, it's a supercomputer, it's a Land-Rover, it slices, it dices—and on reflection, this imaginary nanomachine sounds uncontrolled and dangerous. With enough effort, a do-it-all nanomachine could perhaps be built, but it sounds difficult and there's no good reason to try.

There are advantages to making systems of molecular machinery that can use inexpensive, abundant chemicals, and devices that can operate in nature, but these machines needn't be replicators. A facility like Desert Rose might be designed to use little but electric power from solar panels and molecules from the air, but a setup like this isn't going to slip away. Nanomachines built for cleaning up pollutants and other outdoor tasks could be manufactured in facilities run like Desert Rose and then spread or installed where they're needed.

Extraordinary accidents deserve attention, but with a little care they can be completely avoided. The incentive to build anything resembling a Star Trek-scenario replicator is negligible, even from a military perspective. Any effort toward building such a thing should be seen not as a use of nanotechnology, but as an abuse. Other abuses seem more likely, however, and are quite bad enough.

Chief Danger: Abuse

The chief danger of nanotechnology isn't accidents, but abuse. The safety benefits of nanotechnology, when used with normal care, will free some of our attention to grapple with this far more difficult problem. As Lester Milbrath observes, "Nanotechnologies have such great power that they could be used for evil or environmentally destructive purposes as easily as they could be used for good and environmentally nourishing purposes. This great danger will require a level of political control far beyond that which most nations know how to exercise. We have a prodigious social learning task that we must face."

Thus far, we've focused on how increased abilities can serve constructive ends. Not surprisingly, the potential consequences—with the huge exception of social and economic disruption—are overwhelmingly positive. Inherently clean, well-controlled, inexpensive, superior technologies, when applied with care, can yield far better results than inherently dirty, messy, costly, inferior technologies. This should come as no surprise, but it is only half of the story. The other half is the application of those same superior technologies to destructive ends.

Readers feeling that all this may be too good to be true can breathe a sigh of relief. This problem looks tough.

Cooperative Controls

Molecular manufacturing will lead to more powerful technologies, but our current, crude technology already has world-smashing potential. We have lived with that potential for decades now. In the coming years, we will need to strengthen institutions for maintaining peaceful security.

If most of the political power in the world, and with it most of the police and military power, sees that the course of self interest lies in peace and stability, then solutions seem possible. (The prospect of an arms race in nanotechnology is terrifying and to be avoided at almost any cost. As of this writing, the end of the Cold War offers a better hope of avoiding this nightmare.) James C. Bennett, a high-tech entrepreneur and public policy commentator affiliated with the Center for Constitutional Issues in Technology, explains the goal: "Advanced technologies, particularly as far-ranging a capability as nanotechnology, will create a strong demand for their regulation. The challenge will be to create sufficient controls to prevent the power-hungry from abusing the technologies, without either smothering development or creating an overbearing international regime."

In the coming decades, preventing major abuse of nanotechnology will take the form of regulation, arms control, and antiterrorist activities. In the field of arms control, nanotechnology should present strong motivation for international cooperation and for intimate mutual inspection in the form of joint research programs.

The sheer productive capabilities of molecular manufacturing will make it possible to move from a working weapons prototype to mass production in a matter of days. In a more exotic vein, dangerous nanomachines could be developed, including programmable "germs" (replicating or nonreplicating) for germ warfare. Either development could bring war. With peace looking so profitable and an arms race looking so dangerous, arms control through cooperative development should look attractive. This does not make it easy, or likely.

Terrorism is not an immediate concern. We have lived with nuclear weapons and nerve gas for decades now, and nerve gas, at least, is not difficult to make. As of this writing, no city has been obliterated by terrorists using these means, and no terrorist has even made a credible threat of this sort. The citizens of Hiroshima and Nagasaki, like the Kurds in Iraq, fell victim to nuclear and chemical weapons wielded by governments, not small groups. So long as nanotechnology is technologically more challenging than the simple chemistry of nerve gas, nanoterrorism should not be a primary concern.

To keep dangerous nanotechnologies unavailable, however, will require regulation. If anyone were free to build anything using molecular manufacturing, then someday as the technology base improves and designs become available for more and more nanodevices, someone, somewhere—if only out of sheer spite—would figure out how to combine those nanodevices to make a dangerous replicator and turn it loose. There will almost surely be warning signs, however: In the natural course of events, causes attract protesters before stone-throwers, and produce letter bombs before car bombs. Abuse of nanotechnology is likely to be visible long before it is devastating, and this at least gives some time to try to respond.

Regulatory Tactics

Abuse of this sort can be delayed, perhaps for a long time, by proper regulation. The goal here isn't to make regulations so tight that people will have to violate them to get anything done. This would encourage holdouts, underground work, and disrespect for the law. Instead, the goal is to draw boundaries loosely enough to cause little difficulty for legitimate work, while making dangerous activities very difficult indeed. This is a delicate balance to strike: those fearful of risk naturally try to apply crude and oppressive regulations, and companies naturally try to loosen and avoid regulation entirely. Nonetheless, the problem must be solved, and this seems the best direction to explore.

In one approach, nanomachines could be divided into two classes: *experimental devices* and *approved products*. Approved products could be made widely available through special-purpose molecular manufacturing systems. Thus, once an experimental device had passed regulatory inspection, it could become inexpensive and abundant. In this way, popular demands for a product could be satisfied without anyone needing to break safety rules.

Approved products could include devices like (but superior to) the full range of modern consumer products, ranging from personal supercomputers with 3-D colour displays, through smart construction materials, to running shoes with truly amazing features. The main cost of such goods might be the royalty to the designer. In *Engines of Creation* (the first book to examine this topic), this strategy for producing and distributing approved products is called a "*limited assembler* system."

Note that both approved products and the limited *assemblers* that build them would lack the ability to make copies of themselves, to self-replicate. Ralph Merkle sees this ability as the one to keep an eye on: "Self-replicating systems can and should be appropriately regulated. There seems no need, however, to have any more than normal concerns for devices which cannot replicate. While we might, as with any device, need laws to ensure their appropriate use, they pose no extraordinary problems." For most products, normal medical, commercial, and environmental standards would apply; the regulatory bureaucracies are already in place.

There are great advantages to permitting nearly free experimentation in a new technology, allowing creative people to try ideas without seeking prior approval from a cumbersome committee. Surprisingly, this, too, seems compatible with safety.

In the world of nanotechnology, one cubic micron is a large space, with room enough for millions of components. For many purposes, a few cubic microns would amount to a large laboratory space. To a device on a micron scale, a centimeter is an enormous distance. Surrounding a micron-scale device with a centimeter-thick wall would be like surrounding a human being with a wall kilometers thick, and just as hard to penetrate. Further, a micron-scale device can be incinerated in an instant by something as small as a spark of static electricity. Based on observations like these, *Engines of Creation* outlined the idea of a *sealed assembler lab*, in which a researcher could build anything, even something deliberately designed to be dangerous, and yet be unable to get anything out of the microscopic sealed laboratory except for information.

With a good communications network, a researcher or product developer in Texas could equally easily perform experiments in a remote Maine laboratory run with the security and secrecy of a Swiss bank. A lab would have a responsibility to its customers to keep proprietary work confidential, and a responsibility to regulatory authorities to ensure that nothing but information leaves the laboratory. Researchers could then perform any small-scale experiments they wish. Only approved products, of course, would be built outside the sealed laboratories. While this may not be the best pattern of regulation possible, it does show one way in which freedom of experimentation could be combined with strict regulation of use. By providing a clear separation between legitimate and illegitimate activity, it would help with the difficult problem of identifying and preventing research aimed at damaging ends.

A sensible policy will have to balance the risk of private abuse of technology against the risk of government abuse of technology and regulation. Low-cost manufacturing can make surveillance equipment less expensive. Increased surveillance can reduce some risks in society, but the watchers themselves often aren't very well watched. Placing bounds on surveillance is a challenge for today's citizens as well as tomorrow's, and lessons learned in the past can be applied in the future.

In the long run, it seems wise to assume that someone, somewhere, somehow, will escape the bounds of regulation and arms control and apply molecular-manufacturing capabilities to making novel weapons. If by then we have had several decades of peaceful, responsible, creative development of nanotechnology (or perhaps a few years of help from smart machines), then we may have developed both ecosystem protectors and sophisticated immune machines for medicine. There is good reason to think that distributed technologies of this sort could be adapted and extended to deal with the problem of protecting against novel nanoweaponry. Failure to do so could mean disaster. Nonetheless, building protective systems of this sort will be by far the greatest challenge of any we have discussed. The chief purpose of regulatory tactics like those we have described must be to buy time for those peaceful developments, to maximize the chances that this challenge can be met before time runs out.

Guide It, or Stop It?

Potential accidents richly deserve the attention they will get, and we have confidence that this attention will suffice to make nanotechnology a force for improved human and environmental safety. Abuse is the greater danger, and harder to deal with. When considering a proposed policy, the first question should be, "How will this affect the long-term likelihood of serious abuse?"

Guiding Means Making Many Choices

Guiding a technology is a complex task. It means grappling with myriad decisions regarding which applications are beneficial and which harmful in such complex areas as medicine, the economy, and the environment. It means making such happy choices as which of several good approaches to apply in cleaning up toxic-waste dumps and reversing the greenhouse effect. It also means making more difficult choices in planning ecosystem restoration and environmental modification.

These problems will confront us with a range of choices better than we have today, yet choices that throw values into conflict. Which is a better use for a particular piece of land—the slow restoration of a wilderness ecosystem, or development as a recreational park? Either may be far better than pavement, strip mines, and dumps, but the choices will be controversial.

Likewise, in medicine, we will have a choice of developing many different ways to cure cancers, many different ways to cure heart disease, many different ways to cure AIDS. But the technologies that can be used to rebuild damaged heart muscle could be extended to rebuild muscle and connective tissue structures elsewhere in the body, without the harmful side effects of steroid drugs. The range of choices open to people will be enormous, and again will be cause for great debate.

When a new medical technology is discussed today, a frequent comment is, "This procedure raises ethical questions." This is often taken as a signal to delay its use, neglecting such ethical questions as "Is withholding this lifesaving treatment while we ponder akin to murder?" When a choice raises ethical questions or throws values into conflict, it is time to make an ethical decision or to step aside and let others choose for themselves. Deciding to avoid whatever raised the question is itself a decision—and often ethically indefensible. New technologies will face us with uncomfortable decisions, but so does life itself.

Setting up rules for nanotechnology development will be challenging: finding ways to maximize research freedom while preventing serious abuse and making this stick worldwide is a social challenge of the first rank. Beyond this are decisions regarding rules for its application, and the challenge of maximizing freedom of choice and action while preventing serious abuse, again *worldwide*.

To guide nanotechnology means grappling with a set of decisions that could ultimately remake much of the world—for the better if we are reasonably wise, or for the worse if we are too blundering and incautious. To avoid this responsibility (if we could) would be tempting, yet given the environmental and human stakes, it would, perhaps, be a wrong of historic proportions.

Trying to Stop Means Losing Control

The simplest imaginable approach to "guiding" nanotechnology would be to stop it. The easiest trip to plan is the trip that goes nowhere. This would have a certain appeal, if it were possible. Because of its enormous potential for abuse, nanotechnology has the potential of doing great harm. If we believe that human beings and human institutions are too incompetent to deal with nanotechnology—that they are too likely to turn it to aggressive military use, or too likely to make it freely available to madmen—then the option of stopping the development of nanotechnology may seem attractive indeed. But the ethical question that must guide human actions is not "Would it be better to stop?", but "Would attempts to stop make things better?"

One option is to push forward, emphasizing the need for caution but also the potential for good applications. The promise of medical, economic, and environmental applications, joined with the threat posed by a new arms race, provides a powerful motive for international cooperation. With positive goals and an inclusive stance, international cooperation is a promising strategy; it could provide a basis for guiding the development and application of nanotechnology.

Another option would be to emphasize the downside, to focus debate on potential abuses in support of a campaign to halt development. In following this strategy, an activist group would want to downplay the civilian applications of nanotechnology and emphasize its military applications. Horror stories of potential abuse (including abuses that regulation could easily prevent) would help to make the technology seem strange and dangerous.

This strategy might succeed in suppressing civilian research in many countries, though probably not all. Unfortunately, it would also guarantee funding for classified military research programs in laboratories around the world, even in the most morally honest countries, because of their then-inevitable fear of the consequences if someone else developed nanotechnology first. In a hostile public atmosphere, research would be pushed into secret programs, and in secrecy the prospects for broad international cooperation would disappear. Attempts to stop nanotechnology for fear of a new, unstable arms race become self-fulfilling prophecies. Afterward, the advocates of this view could then say, "We warned you!" as the world slid toward a war they themselves had helped to prepare.

Attempting to stop technological development is a simple but dangerous idea. The greater its success, the greater the polarization it would cause between technology advocates and technology critics. A moderate success would push research out of the public universities and into corporate and military research labs. A greater success would push research out of the corporate laboratories and into heavily classified programs. A truly amazing success would end most of these, leaving the only remaining military programs in the hands of those states with thoroughly repressive governments or alien ideologies. This, presumably, is not how one would prefer nanotechnology to be developed.

The only genuine success would be a total success, and this would mean banning research not only in the United States, and Germany, and France, and the rest of Western Europe, and Japan, and the Soviet

Union, and the People's Republic of China, and Taiwan, but in Korea, South Africa, Iran, Iraq, Israel, Brazil, Argentina, Vietnam, and the part of Colombia controlled by the Medellín Cartel. Later, as computers improve, as chemistry advances, as more and more proximal probe microscopes are built by high school students, total success would require banning kids from tinkering in suburban garages in Pittsburgh.

Competitive pressures are pushing technology toward thorough control of matter, and we have seen that this goal can be reached by many different paths. Preventing one area of research would not prevent the advance, nor would stopping work in one country. When the United States delays drug development through strong regulation by the FDA, drug companies simply switch research overseas, or non-U.S. companies pull ahead. Orbital-launch capability and nuclear weapons capability are other examples. Very seldom has one country given these abilities to another, yet at least eight nations are able to launch satellites to orbit independently, at least seven have detonated nuclear devices, and another two are suspected to be within easy reach of nuclear capability. India and Israel have built bombs and launched satellites, though neither is considered a great power or a leading force in world technology.

Where nanotechnology is concerned, many countries are capable of doing the required research, and more will be in the future. South Korea has both the needed educational levels and the ambition; visitors from the People's Republic of China ask about nanotechnology. A decision at the top directing the resources of a nation could get results almost anywhere. The United States is only gradually being shaken from its illusion that it rules the world of technology. This illusion is a poor basis for decisions and action.

Responsible Action

For all practical purposes, nanotechnology seems inevitable. With work, it can be made beneficial, but only if we exercise ordinary care in avoiding accidents and extraordinary care in preventing abuse.

It's hard to get people to take future technologies seriously. Present-day problems dominate discussions, and ideas about future possibilities take effort to judge. Because of this inertia, broad international regulation of nanotechnology won't be possible until nanotechnology already exists, until people begin to see its results. And then, for regulation to be most effective, researchers and governments in many countries will need to cooperate and be on speaking terms with the technology's critics.

What, then, is the socially responsible course of action, the approach most likely to avoid serious abuse of nanotechnology and most likely to deliver some of its potential benefits? It is, we believe, to point out potential dangers and abuses and how they can be avoided, but also to emphasize the civilian applications in medicine, the environment, and the economy. It is these benefits that provide grounds for advocating open civilian development programs, and for international cooperation that can provide a basis for effective international guidance.

To guide nanotechnology will not be simple. We will be confronted with a range of choices greater than we have faced before in history. It is only by grappling with those choices that we will be able to affect them for the better.

11

POLICY AND PROSPECTS

Although *exploratory engineering* research can show certain technological possibilities, gaining this knowledge can have a paradoxical effect on our *feeling* of knowledge, on our sense of how much we know about the future. It gives us more information, but it can reveal a range of possibilities so vast that we feel as if we know less than we did before.

The prospect of *nanotechnology* and *molecular manufacturing* has this paradoxical effect. It makes certain scenarios—such as a mid-twenty-first-century world of poverty, or choking on pollution caused by massive accumulations of twentieth century-style industry—seem very unlikely indeed. This is useful information in trying to understand our real situation and trying to make sensible plans for the future. And yet the range of new possibilities opened up is broader than we could have imagined before. On the negative side, one can imagine building engines of destruction capable of devastating the world as thoroughly as a nuclear war. On the positive side, one can imagine futures of stable peace with levels of health, wealth, and environmental quality beyond any historical precedent and beyond present expectations.

Within this spectrum of possibilities (and off to its sides) is a range of futures we can't even imagine. Our actions, day by day, are taking us into one of those futures. Not to some future of our present plans or dreams or nightmares, but to a real future, one that will grow from the intended and unintended consequences of our actions, one that we and our descendants will actually have to live in.

Scenarios are useful tools for thinking about the future. They don't represent predictions of what will happen, but instead they present pictures of worlds that one can imagine happening. By looking at these pictures and seeing how they fit together, we can try to get some idea of which events are more likely and which are less likely, and to get some idea of how the choices we make today may affect the shape of things to come.

ORDINARY EXPECTATIONS (1990)

Nanotechnology will have little direct effect on the world until it is well developed, many years from now. The *expectation* of nanotechnology, however, is influencing how people think and act today. Yet even this expectation is still in the early stages of development and will likely have little effect on world affairs for years to come. In sketching scenarios, it seems sensible to begin with the standard worldview, at least for the next few years, and then to look at how nanotechnology and the expectation of nanotechnology might later begin interacting with large-scale developments. As this is being written, old projections of East European, Middle Eastern, and world affairs have recently been upended, and expectations are fairly muddy. Still, one can identify the broad outlines of a conventional-wisdom view of expected events in the coming years and decades:

Technology doesn't change much in the next five years, or indeed in the next fifty. Computer power continues to grow rapidly, but with few important effects. The great challenges of technology are environmental: dealing with greenhouse gases and acid rain and the problems of toxic waste. In parallel, more and more nations climb the ladder of technological capability to such thresholds as the ability to launch satellites, build nuclear weapons, and manufacture computer chips. With the worldwide flow of technical information and the worldwide emphasis on technological development, more and more second-rank countries follow close on the heels of the technological leaders.

Consumer electronics continues to improve, but this leads to a better-entertained population rather than a better-informed one. Exciting announcements like high-temperature superconductors and low-temperature fusion continue to appear, but after hearing cries of "Wolf!" and seeing only puppy dogs and fairy tales, most people discount news of purported breakthroughs.

Even in the thirty-to-fifty-year time frame, most newspaper stories and respected analysts assume there will be little technological change. Fifty-year projections of carbon-dioxide accumulation in the atmosphere assume that most energy will continue to come from fossil fuels. Thirty-year projections of economic crisis due to an aging population and a shrinking work force assume that economic productivity doesn't change greatly.

In terms of productivity and wealth, the United States continues to lose ground relative to the booming economies of Eastern Asia: to Japan, South Korea, Taiwan, and Singapore. In political terms, the Ordinary Expectations scenario is less clear, but expectations seem to run something like this: The breakup of the Eastern bloc and the collapse of communism as a "progressive" ideal lead to a freer and more democratic world. In Eastern Europe, and perhaps in Central Asia, independent countries emerge, each with an industrial base and a population having substantial education in science and technology.

The relative decline of the United States economically and of the Soviet Union militarily loosen some of the ties that today bind the world's democracies to one another. The decreased threat of Soviet military power weakens alliances. As NATO loosens, and as the nations of Europe integrate their economic and political lives, gaps between the United States and Europe grow. As Soviet pressure on Japan weakens, the U.S.-Japanese military alliance weakens and trade frictions loom larger in comparison.

In this environment, protectionist pressures increase. An economic crash grows more likely. A shift from friendly relationships to peaceful hostility becomes an ominous possibility. The rise of multiple, nearly equal centers of economic and technological capability provides incentives for greater integration and cooperation, but also motives for great competition and secrecy.

In the long term, however, limited resources and the costs both of pollution and of pollution controls bring economic growth to a halt in an increasingly impoverished world. Population growth during this time has slowed, but creates great economic and environmental pressures. Resource conflicts escalate into war. The climate has changed irreversibly, the old forests are nearly gone, and extinction has swept a majority of species into nothingness.

Variations on the first five to ten years of the Ordinary Expectations scenario can provide a backdrop for scenarios covering the rise of nanotechnology in, perhaps, the next ten to twenty years:

POLLYANNA TRIUMPHANT

We are living in a world like that of the Ordinary Expectations scenario where, after years of anticipation, primitive but fairly capable *assemblers* have recently been developed. For the first time, the media, the public, and policymakers take the prospect of nanotechnology seriously. It looks very good to them. Technical work has shown that nanotechnology, once developed, can be used in a clean, controlled way, and that it can ultimately displace polluting industries while greatly increasing wealth per capita. The anticipated health benefits are enormous, and after years of a growing death toll from AIDS—only partially stemmed by advances in *molecular medicine*—the public has become very sensitive to the regular reports of human infection by exotic primate *viruses* from Africa. Concern about the stability of Earth's climate and ecosystems has grown as forests have shrunk and weather patterns have changed.

The prospect of breaking out of this cycle is appealing. It is clear that nanotechnology is no danger when in the hands of people of goodwill, and a relatively peaceful decade has allowed many people to forget the existence of other motives.

And so, with miraculously undivided popular support drawn from a grand coalition of environmentalists seeking to replace existing industry, industrialists seeking a more productive technology, health advocates seeking better health care, low-income groups seeking greater wealth, and so on and so forth, companies and governments plunge into nanotechnology with both feet and without reservation.

Development proceeds at a breakneck pace, and everyone who wants to participate in this great venture is welcome. Primitive assemblers are used to build better assemblers, which are used to build yet better assemblers, in laboratories and hobby shops around the world.

Products begin to pour forth. The economy is thrown into turmoil. Military equipment also begins to pour forth, and tensions begin to build. A military research group with more cleverness than sense builds a monster *replicator*, it eats everything, and we all die.

This scenario is absurd, at least in part because published warnings already exist. Since the 1960s, uncritical applause for new technologies has been limited to the now-defunct controlled presses of Eastern Europe (and similar places), and even there the resulting environmental disaster has become a matter for public debate, criticism, and correction.

In the expanding free world of today, the benefits, costs, and dangers of any great new technology will be thoroughly examined, expounded upon, and lied about from many different directions. We may or may not manage to make wise choices as a result. But one thing seems sure: Pollyanna will not triumph, because Pollyanna doesn't have the facts on her side.

Chicken Little Rules the Roost

Again, we are in the world of the Ordinary Expectations scenario, and primitive assemblers have recently been developed. Again, the prospect of nanotechnology is being taken seriously for the first time—but it is somehow portrayed as being just more of the same, but worse. Environmentalists view it not as an alternative to the polluting industries of the twentieth century, but as an extension of human power, and hence of the human ability to do harm. Horror stories of technology gone mad are spun to support this view.

Arms control groups are justifiably alarmed by nanotechnology and emphasize its military applications. Groups seeking arms control via disarmament—and believing in unilateral strategies—work to prevent the development of nanotechnology everywhere they can, that is, everywhere within their political reach. To maximize their political leverage, they portray it as an almost purely military technology of immense and malign power.

Special interest groups in industry see molecular manufacturing as a threat to their business and join the lobbying efforts to prevent it from happening. Unions, neglecting the prospect of greater wealth and leisure for their members, focus instead on possible disruptions in established jobs. They, too, oppose the development of the new technology. As a result, we hear not about how nanotechnology could be used in health care, environmental cleanup, and the manufacture of improved products, but about the insidious threat of tiny, uncontrollable military monster machines that will smash our industry.

After a few years of hearing this, public opinion in the industrial democracies is firmly "against the development of nanotechnology," but this is more a slogan than an enforceable policy. Laws are nevertheless passed to suppress it, and the focus of public debate returns to the old themes of poverty and disease and the newer themes of climatic change and environmental destruction. Solutions seem as distant as ever. No right-thinking person would have anything to do with nanotechnology, so only wrong-thinking people do.

But the initial debate hadn't become serious until assemblers were developed, and research had gone still further before the laws were passed. By then, nanotechnology was just around the corner.

Developing nanotechnology is primarily a matter of tools, just as was developing nuclear weapons. Decades earlier, nuclear-weapons capability had spread from one to two countries in forty-nine months, and to another three in the next fifteen years, despite the requirement for large quantities of exotic materials in each device. By the 1980s, there was already a huge international trade in chemical compounds, and many thousands of chemists who knew how to combine them to make new molecular objects, working not only in university research labs, in corporate research labs, and in civilian and military government research labs, but—as the black market in designer drugs shows—secretly, in criminal research labs.

Even in the 1980s, a scanning tunneling microscope had been built as a high school science-fair project in the United States. There is nothing large-scale or exotic about synthetic chemistry or about precise positioning of *molecules*. And in our scenario, primitive assemblers have already been developed and techniques for constructing them published (as is standard practice) in the open scientific literature.

And so the attempts to suppress the development of nanotechnology succeed only in suppressing the open development of nanotechnology. But governments cannot be sure that other governments are not developing it in secret, and they have now heard so much about its military potential that this is impossible to ignore. Around the world, governments quietly set up secret research programs: some in democracies, others in the remaining authoritarian states.

There are even underground efforts. Once a primitive assembler or even an *AFM*-based *molecular manipulator* is in hand, the remaining challenges are chiefly those of design. In the 1980s, personal computers had become powerful enough to use for designing molecules. In the years since then, computer power has continued its exponential explosion. Peculiar elements of the technoculture join with—pick one: radical anarchists, radical reds, radical greens, or radical racists—in a project aimed at bringing down "the corrupt world order" of governments, of companies, of religions, of human beings, or of nonwhite/nonbrown people. With responsible groups out of the technology race, they see a real chance of finding the leverage needed to change the world.

And so years pass in comparative quiet, with occasional rumors of activity or exposure of a project. Then, from an unexpected direction beyond the reach of democratic control, destructive change breaks loose upon an unprepared world. The sky falls, and Chicken Little is vindicated.

With luck, we will find that this scenario is also absurd. Public debate in the coming years will surely present a more balanced picture of the opportunities and dangers posed by the development of nanotechnology. Thoughtful people with conflicting views will become deeply involved. The impracticability of attempting to suppress technologies of this sort will likely become clear enough to give us a chance of keeping development in the open, in relatively responsible hands.

International Technorivalry

A variant of the Ordinary Expectations scenario has played out for a number of years now. And after years of continuing turbulence, the net result is this: Japanese economic power has grown, with other East Asian economies beginning to close the gap. Their greater investment in long range civilian R&D, with a focus since the late 1980s on engineering molecular systems, has enabled them to take the lead on the path to nanotechnology.

European economic integration and German unification, combined with the pressure of economic and technological competition from the United States and Japan, have turned Europe inward to some extent. Although cultural ties with the United States keep U.S.-European relations on a basically warm basis, hostility between Europe and Japan—already marked in the 1980s—has grown. Europe had long enjoyed great strength in chemistry and basic science, and in the 1980s had led the United States in organizing efforts on *molecular electronics*. This has placed them in a strong position with respect to nanotechnology, behind Japan but ahead of the United States.

The United States remains an enormously productive economy, but the cumulative effects of an educational system that neglects learning and corporations that emphasize quarterly results have made themselves felt. After decades of emphasizing the short term, people now find themselves living in the long term they had neglected. The reaction to U.S. relative economic decline has not been investment and renewal, but rhetoric and hostility directed toward "foreigners," particularly the Japanese.

It is thus an isolated and somewhat defensive Japan that builds the first molecular manipulator and recognizes its long-term potential. The technology is developed in a government-funded research laboratory with cooperation from major Japanese corporations. As the result of increasing tensions, foreign researchers—those still welcome in Japan—were not invited to participate in this particular effort.

A series of committee meetings formalizes a tacit decision made earlier in choosing researchers, and the specifics of this new development are treated as proprietary. Impressive results are announced, stirring pride in Japanese research, but the specifics of the methods involved are kept quiet.

This scarcely delays the diffusion of the basic technology. After the first demonstration, even the most myopic funding agencies support projects with the same goal. A European project had already been started in a French laboratory: it soon succeeds in building an assembler based on somewhat different principles. European researchers follow the Japanese precedent by keeping the details of their techniques as a loosely held secret, in the name of European competitiveness. The United States follows suit a year later in an effort funded by the Department of Defense.

Public life goes on much as before, dominated by the antics of entertainers and politicians, and by tales of the fate of the environment or the Social Security system in a fantasy-future of extrapolated twentieth-century technology. But more and more, in policy circles and in the media, there is serious discussion of nanotechnology and molecular manufacturing—what they mean and what to do about them.

In Japan, second-generation assemblers have begun to turn out small quantities of increasingly sophisticated molecular devices. These are prototypes of commercially useful products: sensors, molecular electronic devices, and scientific instruments; some are immediately useful even at a price of a hundred dollars per molecule. There are plans on the drawing boards for molecular assemblers that could make these devices at prices of less than one-trillionth of a dollar. There are long-term plans (viewed with hope and anticipation) for full-fledged molecular manufacturing able to make almost anything at low cost from common materials.

This is exciting. It promises to at last free Japan from its decades-old dependence on foreign trade, foreign food, foreign raw materials, and foreign politics. By making spaceflight inexpensive and routine, it promises to open the universe to a people cooped up on a crowded archipelago. Investment soars.

Europe leads America but lags behind Japan and looks on Japanese progress with hostility. Europeans, too, share dreams of a powerful technology, and begin a race for the lead. The United States trails, but its huge resources and software expertise help it pick up speed as it joins the race. Other efforts also begin, and though they advance steadily, they cannot keep pace with the great power blocs.

On all sides, the obvious military potential of molecular manufacturing fires military interest, then research and development in both publicly announced and secret programs. Strategists play nanotechnology war games in their minds, in their journals, and on their computers. They come away shaken. The more they look, the more strategies they find that would enable a technologically superior power to make a safe, preemptive move—lethal or nonlethal—against all its opponents. Defenses seem possible in principle, but not in time.

Yet it becomes obvious that molecular manufacturing can provide defenses against lesser technologies. Even the great, mythical leak-proof missile shield looks practical when the defenders have vastly superior technology and a thousandfold cost advantage building military equipment.

No great power seems particularly hostile. By then, all have formally or informally been in a peaceful alliance for many years. Yet there are still memories of war, and the bonds of alliance and military cooperation are weakened by the lack of a common enemy and the growth of economic rivalry. And so squabbles over trade in obsolescing twentieth-century technologies poison cooperation in developing and managing the fresh technologies of the twenty-first century.

There are a thousand reasons to pursue military research and development in these technologies, and nationalistic economic competition helps keep that work secret on a nationalistic basis. Military planners must concern themselves not so much with intentions as with capabilities.

And so a technology developed in an atmosphere of commercial rivalry and secrecy matures in an atmosphere of military rivalry and secrecy. Advanced nanotechnologies arrive in the world not as advances in medicine, or in environmental restoration, or as a basis for new wealth, but as military systems developed in the midst of an accelerating multilateral arms race, with the quiet goal of preemptive use. Negotiations and development run neck and neck, and then ....

Enough Coherence

Again our world is a variant of the Ordinary Expectations scenario, but the international environment is in a healthier condition. Despite trade friction, global economic integration has continued. Europe, the United States, and Japan all have a large stake in each other's well-being, and they recognize it. International military cooperation has continued, in part as a conscious counterweight to conflicts over trade. International cooperation in research has grown, spurred in part by the Japanese desire for closer international ties. The end of the Cold War has made secret military research programs less commonplace. It is in this environment that primitive assemblers are developed, and it doesn't make a great difference who gets there first. As is standard in basic research, groups publish their results in the open literature and compete to impress their colleagues at home and abroad with the brilliance of their achievements.

The arrival of the first assemblers spurs serious debate on nanotechnology and its consequences, and that debate is reasonably open and balanced. It covers military, medical, and environmental consequences, with a major emphasis on how clean, efficient manufacturing can raise the level of wealth and spread it worldwide.

Military analysts consider the impact of molecular manufacturing and its potential products, and concerns are grave, so they undertake classified research programs. But—as usual—secrecy slows communication among researchers: those in the classified programs fall behind their more open colleagues, whose informal information-swapping runs far ahead of the published journals.

Some forces push toward rivalry; others push toward cooperation. A healthy pattern emerges: Those decision makers who take nanotechnology most seriously are precisely those who see the least reason for future international conflict among democratic nations. They no longer anticipate growing conflict over dwindling resources, inequalities of wealth, and global atmospheric pollution. They see what nanotechnology can do for these problems, without anyone taking anything from anyone else. And so, on all sides, those who take nanotechnology most seriously are those most inclined to look for cooperative solutions to the problems it poses. There are exceptions, but the tide of opinion is against them, and their ideas do not dominate policy.

The public debate on nanotechnology grows, and it ranges far and wide. Enthusiasts suggest many wondrous applications for nanotechnology. Some are soon dismissed as being impossible or just plain undesirable. Some are workable improvements on the horrid technologies of the twentieth century; these are developed and applied almost as soon as they become technically possible. The rest are harder to evaluate, but in the course of years of hard work and careful study some of these are developed and adopted, and others are rejected.

At first, some people proposed that nanotechnology be stopped, but they never proposed a credible way to do it. Realists observing the worldwide technological ferment look for other options to deal with the dangers.

The world's industrial democracies, taken together, hold the decisive lead. They have developed mechanisms for coordinating and controlling technologies with military potential by regulating technology transfer and trade. These mechanisms have been developed, exercised, and honed through decades of Cold War experience not only with nuclear and missile technologies, but with a host of high technology products and devices. These mechanisms aren't perfect, but they are useful.

As concerns about international instability mount, the industrial democracies work to improve their teamwork: they reinforce the tradition of free trade and cooperation within the club, and strengthen regulations that block the flow of critical technologies to the world's remaining dictators. As a result of these developments, nanotechnology matures in an atmosphere dominated more by economic cooperation than by military competition. The focus of policy is solidly on civilian applications, with due attention to potential military threats. Trust is reinforced by the automatic "mutual inspection" that is a natural part of cooperative research and development.

Hard decisions remain, and the shouting and the arguments grow louder throughout the world's media. But where the problem is clear, and survival or world well-being are at stake, necessary decisions are made and there is enough international coherence to implement them.

Years pass and technologies mature. Health improves, wealth rises, and the biosphere begins to heal. Despite the turbulence and anguish of change—and despite voices saying, "It was better in the old days," at least for them, and despite real losses—many people of goodwill can look at the world, contemplate the whole, and affirm that this change is, on the whole, a change for the better.

Prospects

Today's knowledge about molecules and matter is enough to give a partial picture of what *molecular machines* and molecular manufacturing will make possible. Even this partial picture shows possibilities that make old views of the twenty-first century thoroughly obsolete.

Science and technology are advancing toward molecular manufacturing along many fronts, in chemistry, physics, biology, and computer science. Motives for continuing range from the medical to the military to the scientific. Research in these directions is already worldwide, and just beginning to focus on the objective of nanotechnology.

Already, it is easy to describe how known devices and principles can be combined to build a primitive device able to guide molecular assembly. Actually doing it will not be so easy—laboratory research never is—but it will be done, and in not too many years.

The first, slow assemblers will lead to products that include better assemblers. Machines able to put molecules together to make molecular machines will lead to a spiral of falling costs and improving quality, ultimately yielding many results that people fervently want: a cleaner environment, an escape from poverty, health care that heals. These benefits will bring disturbing changes and unsettling choices, as new abilities always do. The pace of change may well accelerate, straining the institutions we have evolved to cope with turbulent times.

Molecular-manufacturing capabilities will lend themselves to abuse, and in particular, to the construction of weapons by those seeking power. To minimize the risk of such abuse, we need to develop broad-based international cooperation and regulation. Domestically, this focus seems the best way to avoid polarization between those concerned with solving old problems and those concerned with avoiding new ones. Internationally, it seems the best way to avoid a sickening slide into a new arms race.

As shown by the four scenarios just sketched, public opinion will shape public policy, helping to determine whether these technologies are used for good or for ill. The Afterword will look at today's state of opinion and at what can be done to push in a positive direction. We cannot predict the future, and we cannot predict the consequences of our actions. Nonetheless, what we do will make a difference, and we can begin by trying to avoid every major blunder we can identify. Beyond this, we can try to understand our situation, weigh our basic values, and choose our actions with whatever wisdom we can muster. The choices we make in the coming years will shape a future that stretches beyond our imagining, a future full of danger, yet full of promise. It has always been so.

12

Nanostrategies

A strategy of *personal restraint* (at least in this matter) smacks of simple inaction. But what about a strategy of local political action, of lobbying for laws to suppress certain kinds of research? This would be personal action aimed at enforcing collective inaction. Although it might succeed in suppressing research in a city, a district, a country, or an alliance, this strategy cannot help us guide the leading force instead; it would let some force beyond our control take the lead. A popular movement of this sort can halt research only where the people hold the power, and its greatest possible success would merely open the way for a more repressive state to become the leading force. Where nuclear weapons are concerned, arguments can be made for unilateral disarmament and nonviolent (or at least non-nuclear) resistance. *Nuclear weapons* can be used to smash military establishments and spread terror, but they cannot be used to occupy territory or rule people—not directly. Nuclear weapons have failed to suppress guerrilla warfare and social unrest, so a strategy of disarmament and resistance makes some degree of sense.

The *unilateral suppression* of nanotechnology and *artificial intelligence*, in contrast, would amount to unilateral disarmament in a situation where resistance cannot work. An aggressive state could use these technologies to seize and rule (or exterminate) even a nation of Gandhis, or of armed and dedicated freedom fighters.

This deserves emphasis. Without some novel way to reform the world's oppressive states, simple researchsuppression movements cannot have total success. Without a total success, a major success would mean disaster for the democracies. Even if they got nowhere, efforts of this sort would absorb the work and passion of activists, wasting scarce human resources on a futile strategy. Further, efforts at suppression would alienate concerned researchers, stirring fights between potential allies and wasting further human resources. Its futility and divisiveness make this a strategy to be shunned. Nonetheless, suppression has undeniable appeal. It is simple and direct; "Danger coming? Let's stop it!"

Further, successes in local lobbying efforts promise short-term gratification: "*Danger coming*? We can stop it here and now, for a start!" The start would be a false start, but not everyone will notice. The idea of simple suppression seems likely to seduce many minds. After all, local suppression of local dangers has a long, successful tradition; stopping a local polluter, for example, reduces local pollution. Efforts at local suppression of global dangers will seem similar, however different the effects may be. We will need local organization and political pressure, but they must be built around a workable strategy.

Suppression Agreements

In a more promising approach, we could apply local pressure for the negotiation of a verifiable, worldwide ban. A similar strategy might have a chance in the control of nuclear weapons. But *stopping nanotechnology* and *artificial intelligence* would pose problems of a different order, for at least two reasons. First, these technologies are less well-defined than nuclear weapons: because current nuclear technology demands certain isotopes of rare metals, it is distinct from other activities. It can be defined and (in principle) banned. But modern biochemistry leads in small steps to nanotechnology, and modern computer technology leads in small steps to AI. No line defines a natural stopping point. And since each small advance will bring medical, military, and economic benefits, how could we negotiate a worldwide agreement on where to stop?

Second, these technologies are more potent than nuclear weapons: because reactors and weapons systems are fairly large, inspection could limit the size of a secret force and thus limit its strength. But dangerous replicators will be microscopic, and AI software will be intangible. How could anyone be sure that some laboratory somewhere isn't on the verge of a strategic breakthrough? In the long run, how could anyone even be sure that some hacker in a basement isn't on the verge of a strategic breakthrough? Ordinary verification measures won't work, and this makes negotiation and enforcement of a worldwide ban almost impossible.

Pressure for the right kinds of international agreements will make our path safer, but agreements simply to suppress dangerous advances apparently won't work. Again, local pressure must be part of a workable strategy.

Suppression by Force

If peaceful agreements won't work, one might consider using military force to suppress dangerous advances. But because of verification problems, military pressure alone would not be enough. To suppress advances by force would instead require that one power conquer and occupy hostile powers armed with nuclear weapons-hardly a safe policy. Further, the conquering power would itself be a major technological force with massive military power and a demonstrated willingness to use it. Could this power then be trusted to suppress its own advances? And even if so, could it be trusted to maintain unending, omnipresent vigilance over the whole world? If not, then threats will eventually emerge in secret, and in a world where open work on active shields has been prevented. The likely result would be disaster.

Military strength in the democracies has great benefits, but military strength alone cannot solve our problem. We cannot win safety through a strategy of conquest and research suppression.

These strategies for stopping research—whether through personal inaction, local inaction, negotiated agreement, or world conquest-all seemed doomed to fail. Yet opposition to advances will have a role to play, because we will need selective, intelligently targeted delay to postpone threats until we are prepared for them. Pressure from alert activists will be essential, but to help guide advance, not to halt it.

Non-threatening Policies

If attempts to suppress research in AI and nanotechnology seem futile and dangerous, what of the opposite course—an all-out, unilateral effort? But this too presents problems. We in the democracies probably cannot produce a major strategic breakthrough in perfect secrecy. Too many people would be involved for too many years. Since the Soviet leadership would learn of our efforts, their reaction becomes an obvious concern, and they would surely view a great breakthrough on our part as a great threat. If nanotechnology were being developed as part of a secret military program, their intelligence analysts would fear the development of a subtle but decisive weapon, perhaps based on programmable "*germs*." Depending on the circumstances, our opponents might choose to attack while they still could. It is important that the democracies keep the lead in these technologies, but we will be safest if we can somehow combine this strength with clearly nonthreatening policies.

Balance of Technology

If we follow any of the strategies above we will inevitably stir strong conflict. Attempts to suppress nanotechnology and AI will pit the would be suppressors against the vital interests of researchers, corporations, military establishments, and medical patients. Attempts to gain unilateral advantage through these technologies will pit the cooperating democracies against the vital interests of our opponents. All strategies will stir conflict, but need all strategies split Western societies or the world so badly?

In search of a middle path, we might seek a balance of power based on a *balance of technology*. This would seemingly extend a situation that has kept a measure of peace for four decades. But the key word here is "*seemingly*": the coming breakthroughs will be too abrupt and destabilizing for the old balance to continue. In the past, a country could suffer a technological lag of several years and yet maintain a rough military balance. With swift replicators or advanced AI, though, a lag of a single day could be fatal. A stable balance seems too much to hope for.

Negotiative Development

There is, in principle, a way to ensure a technological balance between the cooperating democracies and Soviet bloc: we could develop the technologies cooperatively, sharing our tools and information. Though this has obvious problems, it is at least somewhat more practical than it may sound.

Is cooperation possible to negotiate? Failed attempts to negotiate effective arms control treaties immediately leap to mind, and cooperation may seem even more complicated and difficult to arrange. But is it? In arms control, each side is attempting to hinder the other's actions; this reinforces their adversarial relationship. Further, it stirs conflicts within each camp between groups that favour arms limitation and groups that exist to build arms. Worse yet, the *negotiations* revolve around words and their meanings, but each side has its own language and an incentive to twist meanings to suit itself.

Cooperation, in contrast, involves both sides working toward a shared goal; this tends to blur the adversarial nature of the relationship. Further, it may lessen the conflicts within each camp, since cooperative efforts would create projects, not destroy them. Finally, both sides discuss their efforts in a shared language—the language of mathematics and diagrams used in science and engineering. Also, cooperation has clear-cut, visible results. In the mid-1970s, the U.S. and U.S.S.R. flew a joint space mission, and until political tensions grew they were laying tentative plans for a joint space station. These were not isolated incidents, in space or on the ground; joint projects and technical exchange have gone on for years. For all its problems, technological cooperation has proved at least as easy as arms control—perhaps even easier, considering the great effort poured into the latter.

Curiously, where AI and nanotechnology are concerned, cooperation and effective arms control would have a basic similarity. To verify an arms control agreement would require constant, intimate inspection of each side's laboratories by the other's experts—a relationship as close as the most thorough cooperation imaginable. But what would cooperation accomplish? It might ensure balance, but balance will not ensure stability. If two gunmen face each other with weapons drawn and fears high, their power is balanced, but the one that shoots first can eliminate the other's threat. A cooperative effort in technology development, unless carefully planned and controlled, would give each side fearsome weapons while providing neither side with a shield. Who could be sure that neither side would find a way to strike a disarming blow with impunity?

And even if one could guarantee this, what about the problem of other powers—and hobbyists, and accidents? As described already a solution to these problems: the development, testing, and construction of active shields. They offer a new remedy for a new problem, and no one has yet suggested a plausible alternative to them. Until someone does, it seems wise to consider how they might be built and whether they might make possible a strategy that can work.

International Agreement

Personal restraint, local action, selective delay, international agreement, unilateral strength, and international cooperation—all these strategies can help us in an effort to develop active shields.

Consider our situation today. The democracies have for decades led the world in most areas of science and technology; we lead today in computer software and biotechnology. Together, we are the leading force. There seems no reason why we cannot maintain that lead and use it. As discussed already, the leading force will be able to use several tactics to handle the assembler breakthrough. These include using sealed assembler labs and limited assemblers, and maintaining secrecy regarding the details of initial assembler development. In the time we buy through these (and other) policies, we can work to develop active shields able to give us permanent protection against the new dangers. This defines a goal. To reach it, a two-part strategy seems best.

The first part involves action within the *cooperating democracies*. We need to maintain a lead that is comfortable enough for us to proceed with caution; if we felt we might lose the race, we might well panic. Proceeding with caution means developing trustworthy institutions for managing both the initial breakthroughs and the development of active shields. The shields we develop, in turn, must be designed to help us secure a future worth living in, a future with room for diversity. The second part of this strategy involves *policies* toward presently *hostile powers*. Here, our aim must be to keep the initiative while minimizing the threat we present. *Technological balance* will not work, and we cannot afford to give up our lead. This leaves strength and leadership as our only real choice, making a *nonthreatening posture* doubly difficult to achieve. Here again we have need for stable, trustworthy institutions: if we can give them a great built-in inertia regarding their objectives, then perhaps even our opponents will have a measure of confidence in them.

To reassure our opponents (and ourselves!) these institutions should be as open as possible, consistent with their mission. We may also manage to build institutions that offer a role for Soviet cooperation. By inviting their participation, even if they refuse the terms we offer, we would offer some measure of reassurance regarding our intentions. If the Soviets were to accept, they would gain a public stake in our joint success.

Still, if the democracies are strong when the breakthroughs approach, and if we avoid threatening any government's control over its own territory, then our opponents will presumably see no advantage in attacking. Thus, we can probably do without cooperation, if necessary.

Defensive Weapons

It may be useful to consider how we might apply the idea of active shields in more conventional fields. Traditionally, defense has required weapons that are also useful for offense. This is one reason why "*defense*" has come to mean "*war-making ability*," and why "*defense*" efforts give opponents reason for fear. Presently proposed space-based defenses will extend this pattern. Almost any defensive system that can destroy attacking missiles could also destroy an opponent's defenses—or enforce a space blockade, preventing an opponent from building defenses in the first place. Such "*defenses*" smell of offense, as seemingly they must, to do their job. And so the arms race gathers itself for another dangerous leap.

Must defense and offense be so nearly inseparable? History makes it seem so. Walls only halt invaders when defended by warriors, but warriors can themselves march off to invade other lands. When we picture a weapon, we naturally picture human hands aiming it and human whim deciding when to fire—and history has taught us to fear the worst.

Yet today, for the first time in history, we have learned how to build defensive systems that are fundamentally different from such weapons. Consider a space-based example. We now can design devices that sense (looks like a thousand missiles have just been launched), assess (this looks like an attempted first strike) and act (try to destroy those missiles!). If a system will fire only at massive flights of missiles, then it cannot be used for offense or a space blockade. Better yet, it could be made incapable of discriminating between attacking sides. Though serving the strategic interests of its builders, it would not be subject to the day-to-day command of anyone's generals. It would just make space a hazardous environment for an attacker's missiles. Like a sea or a mountain range in earlier wars, it would threaten neither side while providing each with some protection against the other.

Though it would use weapons technologies (sensors, trackers, lasers, homing projectiles, and such), this defense wouldn't be a weapons system, because its role would be fundamentally different. Systems of this sort need a distinctive name: they are, in fact, a sort of active shield—a term that can describe any *automated* or *semiautomated system* designed to protect without threatening. By defending both sides while threatening neither, active shields could weaken the cycle of the arms race.

The technical, economic, and strategic issues raised by active shields are complex, and they may or may not be practical in the preassembler era. If they are practical, then there will be several possible approaches to building them. In one approach, the cooperating democracies would build shields unilaterally. To enable other nations to verify what the system will and (more important) won't do, we could allow multilateral inspection of key designs, components, and production steps. We needn't give away all the technologies involved, because know-what isn't the same as know-how. In a different approach, we would build shields jointly, limiting technology transfer to the minimum required for cooperation and verification.

We have more chance of banning space weapons than we do of *banning nanotechnology*, and this might even be the best way to minimize our near-term risks. In choosing a long-term strategy for controlling the arms race, though, we must consider more than the next step. The analysis suggests that traditional arms control approaches, based on negotiating verifiable limitations, cannot cope with nanotechnology. If this is the case, then we need to develop alternative approaches. Active shields—which seem essential, eventually—may offer a new, stabilizing alternative to an arms race in space. By exploring this alternative, we can explore basic issues common to all active shields. If we then develop them, we will gain experience and build institutional arrangements that may later prove essential to our survival.

Active shields are a new option based on new technologies. Making them work will require a creative, interdisciplinary synthesis of ideas in engineering, strategy, and international affairs. They offer fresh choices that may enable us to avoid old impasses. They apparently offer an answer to the ancient problem of protecting without threatening—but not an easy answer.

Power

Nanotechnology and advanced AI may give great power to the leading force—power that can be used to destroy life, or to extend and liberate it. Since we cannot stop these technologies, it seems that we must somehow cope with the emergence of a concentration of power greater than any in history. We will need a suitable system of institutions. To handle complex technologies safely, this system must have ways to judge the relevant facts. To handle great power safely, it must incorporate effective checks and balances, and its purposes and methods must be kept open to public scrutiny. Finally, since it will help us lay the foundations for a new world, it had best be guided by our shared interests, within a framework of sound principles.

We won't start from scratch; we will build on the institutions we have. They are diverse. Not all of our institutions are bureaucracies housed in massive gray buildings; they include such diffuse and lively institutions as the free press, the research community, and activist networks. These decentralized institutions help us control the gray, bureaucratic machines. In part, we face a new version of the ancient and general problem of limiting the abuse of power. This presents no great, fundamental novelty, and the centuries-old principles and institutions of liberal democracy suggest how it may be solved. Democratic governments already have the physical power to blast continents and to seize, imprison, and kill their citizens. But we can live with these capabilities because these governments are fairly tame and stable.

Evil

The coming years will place greater burdens on our institutions. The principles of representative government, free speech, due process, the rule of law, and protection of human rights will remain crucial. To prepare for the new burdens, we will need to extend and reinvigorate these principles and the institutions that support them; protecting free speech regarding technical matters may be crucial. Though we face a great challenge, there is reason to hope that we can meet it. There are also, of course, obvious reasons for doubting that we can meet it. But despair is contagious and obnoxious and leaves people depressed. Besides, despair seems unjustified, despite familiar problems: Evil—are we too wicked to do the right thing? Incompetence—are we too stupid to do the right thing? Sloth—are we too lazy to prepare?

While it would be rash to predict a rosy future, these problems do not seem insurmountable. Democratic governments are big, sloppy, and sometimes responsible for atrocities, yet they do not seem evil, as a whole, though they may contain people who deserve the label. In fact, their leaders gain power largely by appearing to uphold conventional ideas of good. Our chief danger is that policies that seem good may lead to disaster, or that truly good policies won't be found, publicized, and implemented in time to take effect. Democracies suffer more from sloth and incompetence than from evil.

Incompetence

Incompetence will of course be inevitable, but need it be fatal? We human beings are by nature stupid and ignorant, yet we sometimes manage to combine our scraps of competence and knowledge to achieve great things. No one knew how to get to the Moon, and no one ever learned, yet a dozen people have walked its surface. We have succeeded in technical matters because we have learned to build institutions that draw many people together to generate and test ideas. These institutions gain reliability through redundancy, and the quality of their results depends largely on how much we care and how hard we work. When we focus enough attention and resources on reliability, we often succeed. This is why the Moon missions succeeded without a fatality in space, and why no nuclear weapon has ever been launched or detonated by accident. And this is why we may manage to handle nanotechnology and advanced AI with sufficient care to ensure a competent job. Erratic people of limited competence can join to form stable, competent institutions.

Sloth

Sloth—intellectual, moral, and physical—seems perhaps our greatest danger. We can only meet great challenges with great effort. Will enough people make enough effort? No one can say, because no one can speak for everyone else. But success will not require a sudden, universal enlightenment and mobilization. It will require only that a growing community of people strive to develop, publicize, and implement workable solutions—and that they have a good and growing measure of success. This is not so implausible. Concern about technology has become widespread, as has the idea that accelerating change will demand better foresight. Sloth will not snare everyone, and misguided thinkers will not misdirect everyone's effort. Deadly pseudo-solutions (such as blocking research) will lose the battle of ideas if enough people debunk them. And though we face a great challenge, success will make possible the fulfillment of great dreams. Great hopes and fears may stir enough people to enable the human race to win through. Passionate concern and action will not be enough; we will also need sound policies. This will require more than good intentions and clear goals: we must also trace the factual connections in the world that will relate what we do to what we get.

As we approach a technological crisis of unprecedented complexity, it makes sense to try to improve our institutions for judging important technical facts. How else can we guide the leading force and minimize the threat of terminal incompetence? Institutions evolve. To evolve better fact-finding institutions, we can copy, adapt, and extend our past successes. These include the free press, the scientific community, and the courts. All have their virtues, and some of these virtues can be combined.

13

WORLDWIDE TRANSFORMATION

A worldwide transformation which can, if we succeed, bring abundance and long life to all who wish them. And this is a prospect that quite naturally stirs dreams of *utopia*. A standard-issue utopia, as everyone knows, would be static, boring, and dreadful—in fact, it would be no utopia at all. Yet again and again utopian dreams have changed history, whether for good or ill. Dangerous dreams have led people to kill in the name of love, and to enslave in the name of brotherhood. All too often the dream has been impossible and the attempt to achieve it has been disastrous.

We need useful dreams to guide our actions. A useful dream must show us a possible, desirable goal, and steps toward that goal must bring positive results. To help us cooperate in guiding the technology race, we will need goals that appeal to people with differing dreams—but what goals could serve? It seems that they must hold room for diversity. Likewise, what goals chosen today, so near the dawn of intelligence, could prove worthy of the future's potential? It seems that they must hold room for progress.

Only one sort of future seems broad enough to have broad appeal: an open future of liberty, diversity, and peace. With room for the pursuit of many different dreams, an open future will appeal to many different people. Grander schemes, such as establishing a uniform world order, seem more dangerous. If "*one world, or none*" means imposing a single social system on a world of hostile nuclear powers, then it seems a recipe for disaster. "*Many worlds, or none*" seems our real choice, if we can develop active shields to secure peace. We may be able to do so. Using automated engineering systems of the sort already described, we will be able to explore the limits of the possible at a million times human speed. We will thus be able to outline the ultimate limits to the technology race, including the arms race.

With shields based on that knowledge, it seems that we would be able to secure a stable, durable peace. Advancing technology need not push the world into any single mold. Many people once feared that ever larger machines and ever larger organizations would dominate our future, crushing diversity and human choice. Indeed, machines can grow bigger, and some may. Organizations can grow bigger, and some may. But stinking, clanking machines and huge bureaucracies have already begun to seem old-fashioned compared to microcircuits, biotechnology, and fluid organizations.

We now can see the outlines of a higher technology on a human scale, of a world with machines that don't clank, chemical plants that don't stink, and production systems that don't use people as cogs. Nanotechnology shows that advances can bring a different style of technology. Assemblers and AI will let us create complex products without complex organizations. Active shields will let us secure peace without a massive military-industrial complex. These technologies will broaden our choices by loosening our constraints, making room for greater diversity and independence. Establishing an era of universal wealth will require only that the vast, unclaimed resources of space someday be divided in a way that gives everyone a significant share.

Advancing Nanotechnology

Advancing technology may end or extend life, but it can also change its quality. Products based on nanotechnology will permeate the daily lives of people who choose to use them. Some consequences will be trivial; others may be profound. Some products will have effects as ordinary as simplifying housekeeping (and as substantial as reducing the causes of domestic quarrels). It should be no great trick, for example, to make everything from dishes to carpets self-cleaning, and household air permanently fresh. For properly designed nanomachines, dirt would be food.

Other systems based on nanotechnology could produce fresh food—genuine meat, grain, vegetables, and so forth—in the home, year round. These foods result from cells growing in certain patterns in plants and animals; cells can be coaxed to grow in these same patterns elsewhere. Home food growers will let people eat ordinary diets without killing anything. The animal rights movement (the forerunner of a movement to protect all conscious, feeling entities?) will be strengthened accordingly. Nanotechnology will make possible high-resolution screens that project different images to each eye; the result will be three-dimensional television so real that the screen seems like a window into another world. Screens of this sort could line the helmet of a suit much like the spacesuit. The suit itself, rather than being programmed to transmit forces and textures from outside, could instead apply to the skin forces and textures defined by a complex, interactive program. A suit and helmet combination of this sort could simulate most of the sights and sensations of an entire environment, whether real or imaginary. Nanotechnology will make possible vivid art forms and fantasy worlds far more absorbing than any book, game, or movie. *Advanced technologies* will make possible a whole world of products that make modern conveniences seem inconvenient and dangerous. Why shouldn't objects be light, flexible, durable, and cooperative? Why shouldn't walls look like whatever we want, and transmit only the sounds we want to hear? And why should buildings and cars ever crush or roast their occupants? For those who wish, the environment of daily life can resemble some of the wilder descriptions found in science fiction.

Science Dreams

Toward many extremes lie *science fiction dreams*, for those who want to live them. They range from homes that cooperate with us for our comfort to opportunities for toil on distant planets. Science fiction authors have imagined many things, some possible and others in flat contradiction to known natural law. Some dreamed of spaceflight, and spaceflight came. Some dreamed of robots, and robots came. Some dreamed of cheap spaceflight and intelligent robots, and these too are coming. Other dreams seem possible. Authors have written of the direct sharing of thoughts and emotions from mind to mind. Nanotechnology seems likely to make possible some form of this by linking neural structures via *transducers* and *electromagnetic signals*. Though limited to the speed of light, this sort of telepathy seems as possible as telephony.

Starships, *space settlements*, and *intelligent machines* will all become possible. All this lies outside the skin, yet authors have written also of transformations within the skin; these, too, will become possible. Becoming completely healthy in body and brain is one form of change, yet some people will want more. They will seek changes on a level deeper than mere health and wealth. Some will seek fulfillment in the world of the spirit; though that quest lies beyond the scope of crude material technology, new physical possibilities will provide new starting points and time enough to try. The technology underlying cell repair systems will allow people to change their bodies in ways that range from the trivial to the amazing to the bizarre. Such changes have few obvious limits. Some people may shed human form as a caterpillar transforms itself to take to the air; others may bring plain humanity to a new perfection. Some people will simply cure their warts, ignore the new butterflies, and go fishing. Biostasis opens travel into the future, since it can make years pass in an eye blink. The jaded may seek the novelties of a more distant future, perhaps awaiting slowly maturing developments in the arts or society, or the mapping of the worlds of the galaxy. If so, they will choose sleep, passing from age to age in search of a time that suits them. Strange futures lie open, holding worlds beyond our imagining.

ADVANCED SIMPLICITY

E. F. Schumacher, author of Small Is Beautiful, wrote: "I have no doubt that it is possible to give a new direction to technological development, a direction that shall lead it back to the real needs of man, and that also means: to the actual size of man. Man is small, and therefore small is beautiful." Schumacher was not writing of nanotechnology, but could such an advanced technology be part of a simpler life on a human scale? In prehistoric times, people used two sorts of materials: the products of natural bulk processes (such as stone, water, air, and clay) and the products of natural molecular machinery (such as bone, wood, hide, and wool). Today we use these same materials and complex bulk processes to make the products of our global industrial civilization. If technological systems have grown past human scale, our bulk technology and stupid machines are largely to blame: to make systems complex, we have had to make them big. To make them capable, we have had to fill them with people. The resulting system now sprawls across continents, entangling people in a global web. It has offered escape from the toil of subsistence farming, lengthening lives and bringing wealth, but at a cost that some consider too high.

Nanotechnology will open new choices. Self-replicating systems will be able to provide food, health care, shelter, and other necessities. They will accomplish this without bureaucracies or large factories. Small, selfsufficient communities can reap the benefits.

One test of the freedom a technology offers is whether it frees people to return to primitive ways of life. Modern technology fails this test; molecular technology succeeds. As a test case, imagine returning to a stoneage style of life—not by simply ignoring molecular technology, but while using it.

Stone-age villagers lacking modern education wouldn't understand molecular machinery, but this matters little. Since ancient times, villagers have used the molecular machinery of yeast, seeds, and goats without molecularlevel understanding. If such complex and unruly things as goats suit primitive ways of life, then other forms of molecular machinery will surely qualify. Living things show that the machinery inside a self-replicating system can be ignored in a way that the machinery inside an automobile cannot. Thus a group could raise novel "*plants*" and "*animals*" to ease the harsh edges of existence, and yet live a basically stone-age life. They could even limit themselves to ordinary plants and animals, engineered only by millennia of selective breeding.

With possibilities so broad, some people may even choose to live as we do today: with traffic noise, smells, and danger; with pitted teeth and whining drills; with aching joints and sagging skin; with joys off-set by fear, toil, and approaching death. But unless they were brainwashed to obliterate their knowledge of better choices, how many people would willingly resign themselves to such lives? Perhaps a few.

Can one imagine living an ordinary life in a space settlement? A settlement would be large, complex, and located in space—but the Earth is also large, complex, and located in space. Worlds in space could be as selfmaintaining as the Earth and as big as a continent, flooded with sunlight, filled with air, and holding a biocylinder if not a biosphere.

Worlds in space need not be products of direct human design. Underlying much of the beauty of nature is a certain kind of disorderly order. The veins on a leaf, the branches on a tree, the landforms in a watershed—all these have a freedom of form within patterns that resemble what mathematicians call "*fractals.*" Lands in space need not be modeled on golf courses and suburban lots. Some will be shaped with the aid of computers programmed to reflect a deep knowledge of natural processes, melding human purpose with a natural quality that no human mind and hand can directly produce. Mountains and valleys in lands much like wilderness will mirror the shapes of dream-rock and dream-soil, sculpted by dream-ages of electronic water. Worlds in space will be worlds.

Scope for Dreams

This, then, is the size of the future's promise. Though limits to growth will remain, we will be able to harvest solar power a trillion times greater than all the power now put to human use. From the resources of our solar system, we will be able to create land area a million times that of Earth. With assemblers, automated engineering, and the resources of space we can rapidly gain wealth of a quantity and quality beyond past dreams. Ultimate limits to lifespan will remain, but cell repair technology will make perfect health and indefinitely long lives possible for everyone. These advances will bring new engines of destruction, but they will also make possible active shields and arms control systems able to stabilize peace.

In short, we have a chance at a future with room enough for many worlds and many choices, and with time enough to explore them. A tamed technology can stretch our limits, making the shape of technology pinch the shape of humanity less. In an open future of wealth, room, and diversity, groups will be free to form almost any society they wish, free to fail or set a shining example for the world. Unless your dreams demand that you dominate everyone else, chances are that other people will wish to share them. If so, then you and those others may choose to get together to shape a new world. If a promising start fails—if it solves too many problems or too fewthen you will be able to try again. Our problem today is not to plan or build utopias but to seek a chance to try.

Proceedings

We may fail. Replicating assemblers and AI will bring problems of unprecedented complexity, and they threaten to arrive with unprecedented abruptness. We cannot wait for a fatal error and then decide what to do about it; we must use these new technologies to build active shields before the threats are loosed. Fortunately for our chances, the approaching breakthroughs will become steadily more obvious. They will eventually seize public attention, guaranteeing at least a measure of foresight. But the earlier we start planning, the better our chances. The world will soon become hospitable to memes that purport to describe sound policy for the assembler and AI breakthroughs. Such memes will then spread and become entrenched, whether they deserve to be or not. Our chances will be better if, when that time comes, a sound set of ideas has been hammered out and has begun to spread—public opinion and public policy will then be more likely to jump in a sensible direction when the crisis nears.

This situation makes careful discussion and public education important right now. Guiding technology will also require new institutions, and institutions do not evolve overnight. This makes work on hypertext and fact forums important right now. If they are ready to use, then they too will grow more popular as the crisis nears.

Despite the broad appeal of an open future, some people will oppose it. The power-hungry, the intolerant idealists, and a handful of sheer people-haters will find the prospect of freedom and diversity repugnant. The question is, will they shape public policy? Governments will inevitably subsidize, delay, classify, manage, bungle, or guide the coming breakthroughs. The cooperating democracies may make a fatal error, but if they do, it will likely be the result of public confusion about which policies will have which consequences. There will be genuine opposition to an open future, based on differing (and often unstated) values and goals, but there will be far greater disagreements over specific proposals, based on differing beliefs regarding matters of fact. And though many disagreements will stem from differences of judgment, many will inevitably stem from simple ignorance. Even solid, well-established facts will at first remain little known.

Worse, the prospect of technologies as fundamental as assemblers, AI, and cell repair machines must inevitably upset many old, entrenched ideas at once. This will cause conflicts in people's minds (I know; I've experienced some of them). In some minds, these conflicts will trigger the reject-the-new reflex that serves as humankind's most basic mental immune system. This reflex will make ignorance tenacious.

Worse yet, the spread of half-truths will also cause harm. To function properly, some memes must be linked to others. If the idea of nanotechnology were free from the idea of its danger, then nanotechnology would be a greater danger than it already is. But in a world grown wary of technology, this threat seems slight. Yet other idea fragments will spread, sowing misunderstanding and conflict.

The fact forum idea, when discussed without the distinctions among facts, values, and policies, will sound technocratic. Active shields, when proposed without mention of *hypertext* or fact forums, may seem impossible to trust. The danger and inevitability of nanotechnology, to those ignorant of active shields, will bring despair. The danger of nanotechnology, when its inevitability is not understood, will spur futile local efforts to stop its global advance. Active shields, when not motivated by the eventual requirements for controlling molecular technology, will strike some people as too much trouble. When called "*defense projects*," without distinction between defense and offense, shields will strike some people as threats to peace.

Likewise the idea of long life, when unaccompanied by the expectation of abundance and new frontiers, will seem perverse. Abundance, when imagined without space development or controlled replicators, will sound environmentally damaging. The idea of biostasis, to those who know nothing about cell repair and confuse expiration with dissolution, will sound absurd. Unless they are held together by book covers or hypertext links, ideas will tend to split up as they travel. We need to develop and spread an understanding of the future as a whole, as a system of interlocking dangers and opportunities. This calls for the effort of many minds.

The incentive to study and spread the needed information will be strong enough: the issues are fascinating and important, and many people will want their friends, families, and colleagues to join in considering what lies ahead. If we push in the right directions—learning, teaching, arguing, shifting directions, and pushing further—then we may yet steer the technology race toward a future with room enough for our dreams. Eons of evolution and millennia of history have prepared this challenge and quietly presented it to our generation.

The coming years will bring the greatest turning point in the history of life on Earth. To guide life and civilization through this transition is the great task of our time. If we succeed (and if you survive) then you may be honored with endless questions from pesky greatgrandchildren: "What was it like when you were a kid, back before the Breakthrough?" and "What was it like growing old?" and "What did you think when you heard the Breakthrough was coming?" and "What did you do then?" By your answers you will tell once more the tale of how the future was won.

14

NANO PHARMACEUTICAL

What if doctors could search out and destroy the very first cancer cells that would otherwise have caused a tumor to develop in the body? What if a broken part of a cell could be removed and replaced with a miniature biological machine? What if pumps the size of molecules could be implanted to deliver life-saving medicines precisely when and where they are needed? These scenarios may sound unbelievable, but they are the long-term goals of the NIH Roadmap's Nanomedicine initiative that we anticipate will yield medical benefits as early as 10 years from now.

Nanomedicine, an offshoot of nanotechnology, refers to highly specific medical intervention at the molecular scale for curing disease or repairing damaged tissues, such as bone, muscle, or nerve. A nanometer is one-billionth of a meter, too small to be seen with a conventional lab microscope. It is at this size scale–about 100 nanometers or less—that biological molecules and structures inside living cells operate. Nanotechnology involves the creation and use of materials and devices at the level of molecules and atoms. Research in nanotechnology began with applications outside of medicine and is based on discoveries in physics and chemistry. This is because it is essential to understand the physical and chemical properties of molecules or complexes of molecules in order to control them. The same holds true for the molecules and structures inside living tissues.

Researchers have developed powerful tools to extensively categorize the parts of cells in vivid detail, and we know a great deal about how these intracellular structures operate. Yet, scientists have still not been able to answer questions such as, "How many?" "How big?"

and "How fast?" These questions must be addressed in order to build "nano" structures or "nano" machines that are compatible with living tissues and can safely operate inside the body. Once these questions are answered, we will design better diagnostic tools and engineer structures for more specific treatments of disease and repair of tissues.

The centers will be staffed by highly multidisciplinary scientific teams including biologists, physicians, mathematicians, engineers and computer scientists. Research conducted over the first few years will be directed toward gathering extensive information about the physical properties of intracellular structures that will inform us about how biology's molecular machines are built.

As this catalogue of the interactions between molecules and larger structures develops, patterns will emerge, and we will have a greater understanding of the intricate operations of molecular structures, processes, and networks inside living cells. Mapping these networks and understanding how they change over time is crucial to help us understand nature's rules of biological design that, in turn, will enable researchers to use this information to correct biological defects in unhealthy cells. This knowledge will lead to the development of new tools that will work at the "nano" scale and allow scientists to build synthetic biological devices, such as tiny sensors to scan for the presence of infectious agents or metabolic imbalances that could spell trouble for the body, and miniature devices to destroy the infectious agents or fix the "broken" parts in the cells. This initiative is an important component of the NIH Roadmap endeavor because these tools will be developed and applied, not just for a single disease or particular type of cell, but for a wide range of tissues and diseases.

Molecular Body

Our bodies are filled with intricate, active molecular structures. When those structures are damaged, health suffers. Modern medicine can affect the workings of the body in many ways, but from a molecular viewpoint it remains crude indeed. Molecular manufacturing can construct a range of medical instruments and devices with far greater abilities. The body is an enormously complex world of molecules. With nanotechnology to help, we can learn to repair it. To understand what nanotechnology can do for medicine, we need a picture of the body from a molecular perspective.

The human body can be seen as a workyard, construction site, and battleground for molecular machines. It works remarkably well, using systems so complex that medical science still doesn't understand many of them. Failures, though, are all too common. Molecular machines do the daily work of the body. When we chew and swallow, muscles drive our motions. Muscle fibers contain bundles of molecular fibers that shorten by sliding past one another. In the stomach and intestines, the molecular machines we call digestive enzymes break down the complex molecules in foods, forming smaller molecules for use as fuel or as building blocks. Molecular devices in the lining of the digestive tract carry useful molecules to the bloodstream.

Meanwhile, in the lungs, molecular storage devices called hemoglobin molecules pick up oxygen. Driven by molecular fibers, the heart pumps blood laden with fuel and oxygen to cells. In the muscles, fuel and oxygen drive contraction based on sliding molecular fibers. In the brain, they drive the molecular pumps that charge nerve cells for action. In the liver, they drive molecular machines that build and break down a whole host of molecules. And so the story continues through all the work of the body.

Yet each of these functions sometimes fails, whether through damage or inborn defect. In growing, healing, and renewing tissue, the body is a construction site. Cells take building materials from the bloodstream. Molecular machinery programmed by the cell's genes uses these materials to build biological structures: to lay down bone and collagen, to build whole new cells, to renew skin, and to heal wounds. With the exception of tooth fillings and other artificial implants, everything in the human body is constructed by molecular machines. These molecular machines build molecules, including more molecular machines. They clear away structures that are old or out of place, sometimes using machinery like digestive enzymes to take structures apart.

During tissue construction, whole cells move about, amoeba-like: extending part of themselves forward, attaching, pulling their material along, and letting go of the former attachment site behind them. Individual cells contain a dynamic pattern of molecules made of components that can break down but can also be replaced. Some molecular machines in the cell specialize in digesting molecules that show signs of damage, allowing them to be replaced by fresh molecules made according to genetic instructions. Components inside cells form their complex patterns by self-assembly, that is, by sticking to the proper partners.

Failures in construction increase as we age. Teeth wear and crack and aren't replaced; hair follicles stop working; skin sags and wrinkles.

The eye's shape becomes more rigid, ruining close vision. Younger bodies can knit together broken bones quickly, making them stronger than before, but osteoporosis can make older bones so fragile that they break under minor stress.

Sometimes construction is botched from the beginning due to a missing or defective genetic code. In hemophilia, bleeding fails to stop due to the lack of blood clotting factor. Construction of muscle tissue is disrupted in 1 in 3,300 male births by muscular dystrophy, in which muscles are gradually replaced by scar tissue and fat; the molecule "dystrophin" is missing. Sickle cell anemia results from abnormal hemoglobin molecules.

Paraplegics and quadriplegics know that some parts of the body don't heal well. The spinal cord is an extreme—and extremely serious—case, but scarring and improper regrowth of tissues result from many accidents. If tissues always regrew properly, injury would do no permanent physical damage.

Immunity in the Body

Assaults from outside the body turn it into a battlefield where the aggressors sometimes get the upper hand. From parasitic worms to protozoa to fungi to bacteria to viruses, organisms of many kinds have learned to live by entering the body and using their molecular machinery to build more of themselves from the body's building blocks. To meet this onslaught, the body musters the defenses of the immune system—an armada of its own molecular machines. Your body's own amoeba-like white blood cells patrol the bloodstream and move out into tissues, threading their way between other cells, searching for invaders.

How can the immune system distinguish the hundreds of kinds of cells that should be in the body from the invading cells and viruses that shouldn't? This has been the central question of the complex science of immunology. The answer, as yet only partially understood, involves a complex interplay of molecules that recognize other molecules by sticking to them in a selective fashion. These include free-floating antibodies—which are a bit like bumbling guided missiles—and similar molecules that are bound to the surface of white blood cells and other cells of the immune system, enabling them to recognize foreign surfaces on contact.

This system makes life possible, defending our bodies from the fate of meat left at room temperature. Still, it lets us down in two basic ways.

First, the immune system does not respond to all invaders, or responds inadequately. Malaria, tuberculosis, herpes, and AIDS all have their strategies for evading destruction. Cancer is a special case in which the invaders are altered cells of the body itself, sometimes successfully masquerading as healthy cells and escaping detection.

Second, the immune system sometimes overresponds, attacking cells that should be left alone. Certain kinds of arthritis, as well as lupus and rheumatic fever, are caused by this mistake. Between attacking when it shouldn't and not attacking when it should, the immune system often fails, causing suffering and death.

Contemporary Medicine

When the body's working, building, and battling goes awry, we turn to medicine for diagnosis and treatment. Today's methods, though, have obvious shortcomings. Diagnostic procedures vary widely, from asking a patient questions, through looking at X-ray shadows, through exploratory surgery and the microscopic and chemical analysis of materials from the body. Doctors can diagnose many ills, but others remain mysteries. Even a diagnosis does not imply understanding: doctors could diagnose infections before they knew about germs, and today can diagnose many syndromes with unknown causes. After years of experimentation and untold loss of life, they can even treat what they don't understand—a drug may help, though no one knows why.

Leaving aside such therapies as heating, massaging, irradiating, and so forth, the two main forms of treatment are surgery and drugs. From a molecular perspective, neither is sophisticated. Surgery is a direct, manual approach to fixing the body, now practiced by highly trained specialists. Surgeons sew together torn tissues and skin to enable healing, cut out cancer, clear out clogged arteries, and even install pacemakers and replacement organs. It's direct, but it can be dangerous: anesthetics, infections, organ rejection, and missed cancer cells can all cause failure. Surgeons lack fine-scale control. The body works by means of molecular machines, most working inside cells. Surgeons can see neither molecules nor cells, and can repair neither.

Drug therapies affect the body at the molecular level Some therapies—like insulin for diabetics—provide materials the body lacks. Most like antibiotics for infections-introduce materials no human body produces. A drug consists of small molecules; in our simulated molecular world, many would fit in the palm of your hand. These molecules are dumped into the body (sometimes directed to a particular region by a needle or the like), where they mix and wander through blood and tissue. They typically bump into other molecules of all sorts in all places, but only stick to and affect molecules of certain kinds.

Antibiotics like penicillin are selective poisons. They stick to molecular machines in bacteria and jam them, thus fighting infection. Viruses are a harder case because they are simpler and have fewer vulnerable molecular machines. Worms, fungi, and protozoa are also difficult, because their molecular machines are more like those found in the human body, and hence harder to jam selectively. Cancer is the most difficult of all. Cancerous growths consist of human cells, and attempts to poison the cancer cells typically poison the rest of the patient as well. Other drug molccules bind to molecules in the human body and modify their behaviour. Some decrease the secretion of stomach acid, others stimulate the kidneys, many affect the molecular dynamics of the brain. Designing drug molecules to bind to specific targets is a growth industry today, and provides one of the many short-term payoffs that is spurring developments in molecular engineering.

Current medicine is limited both by its understanding and by its tools. In many ways, it is still more an art than a science. Mark Pearson of Du Font points out, "In some areas, medicine has become much more scientific, and in others not much at all. We're still short of what we would consider a reasonable scientific level. Many people don't realize that we just don't know fundamentally how things work. It's like having an automobile, and hoping that by taking things apart, we'll understand something of how they operate. We know there's an engine in the front and we know it's under the hood, we have an idea that it's big and heavy, but we don't really see the rings that allow pistons to slide in the block. We don't even understand that controlled explosions are responsible for providing the energy that drives the machine."

Better tools could provide both better knowledge and better ways to apply that knowledge for healing. Today's surgery can rearrange blood vessels, but is far too coarse to rearrange or repair cells. Today's drug therapies can target some specific molecules, but only some, and only on the basis of type. Doctors today can't affect molecules in one cell while leaving identical molecules in a neighbouring cell untouched because medicine today cannot apply surgical control to the molecular level.

Nanotechnology and Medicine

Developments in nanotechnology will result in improved medical sensors. As protein chemist Bill DeGrado notes, "Probably the first use you may see would be in diagnostics: being able to take a tiny amount of blood from somebody, just a pinprick, and diagnose for a hundred different things. Biological systems are already able to do that, and I think we should be able to design molecules or assemblies of molecules that mimic the biological system."

In the longer term, though, the story of nanotechnology in medicine will be the story of extending surgical control to the molecular level. The easiest applications will be aids to the immune system, which selectively attack invaders outside tissues. More difficult applications will require that medical nanomachines mimic white blood cells by entering tissues to interact with their cells. Further applications will involve the complexities of molecular-level surgery on individual cells.

As we look at how to solve various problems, you'll notice that some that look difficult today will become easy, while others that might seem easier turn out to be more difficult. The seeming difficulty of treating disorders is always changing: Once polio was frequent and incurable, today it is easily prevented. Syphilis once caused steady physical decline leading to insanity and death; now it is cured with a shot. Athlete's foot has never been seen as a great scourge, yet it remains hard to cure. Likewise with the common cold. This pattern will continue: Deadly diseases may be easily dealt with, while minor ills remain incurable, or vice versa. As we will see, a mature nanotechnology-based medicine will be able to deal with almost any physical problem, but the order of difficulty may be surprising. Nature cares nothing for our sense of appropriateness. Horribleness and difficulty just aren't the same thing.

One approach to nanomedicine would make use of microscopic mobile devices built using molecular-manufacturing equipment. These would resemble the ecosystem protectors and mobile cleanup machines discussed in this chapter. Like them, they would either be biodegradable, self-collecting, or collected by something else once they were done working. Like them, they would be more difficult to develop than simple, fixed-location nanomachines, yet clearly feasible and useful. Development will start with the simpler applications, so let's begin by looking at what can be done without entering living tissues.

The skin is the body's largest organ, and its exposed position subjects it to a lot of abuse. This exposed position, though, also makes it easier to treat. Among the earlier applications of molecular manufacturing may be those popular, quasimedical products, cosmetics. A cream packed with nanomachines could do a better and more selective job of cleaning than any product can today. It could remove the right amount of dead skin, remove excess oils, add missing oils, apply the right amounts of natural moisturizing compounds, and even achieve the elusive goal of "deep pore cleaning" by actually reaching down into pores and cleaning them out. The cream could be a smart material with smooth-on, peel-off convenience.

The mouth, teeth, and gums are amazingly troublesome. Today, daily dental care is an endless cycle of brushing and flossing, of losing ground to tooth decay and gum disease as slowly as possible. A mouthwash full of smart nanomachines could do all that brushing and flossing do and more, and with far less effort--making it more likely to be used.

This mouthwash would identify and destroy pathogenic bacteria while allowing the harmless flora of the mouth to flourish in a healthy ecosystem. Further, the devices would identify particles of food, plaque, or tartar, and lift them from teeth to be rinsed away. Being suspended in liquid and able to swim about, devices would be able to reach surfaces beyond reach of toothbrush bristles or the fibers of floss. As short-lifetime medical nanodevices, they could be built to last only a few minutes in the body before falling apart into materials of the sort found in foods (such as fiber). With this sort of daily dental care from an early age, tooth decay and gum disease would likely never arise. If under way, they would be greatly lessened.

Going beyond this superficial treatment would involve moving among and modifying cells. Let's consider what can be done with this treatment inside the body, but outside the body's tissues. The bloodstream carries everything from nutrients to immune-system cells, with chemical signals and infectious organisms besides.

Here, it is useful to think in terms of medical nanomachines that resemble small submarines. Each of these is large enough to carry a nanocomputer as powerful as a mid-1980s mainframe, along with a huge database (a billion bytes), a complete set of instruments for identifying biological surfaces, and tools for clobbering viruses, bacteria, and other invaders. Immune cells, as we've seen, travel through the bloodstream checking surfaces for foreignness and—when working properly—attacking and eliminating what should not be there. These *immune machines* would do both more and less. With their onboard sensors and computers, they will be able to react to the same molecular signals that the immune system does, but with greater discrimination. Before being sent into the body on their search-and-destroy mission, they could be programmed with a set of characteristics that lets them clearly distinguish their targets from everything else. The body's immune system can respond only to invading organisms that had been encountered by that individual's body. Immune machines, however, could be programmed to respond to anything that had been encountered by world medicine.

Immune machines can be designed for use in the bloodstream or the digestive tract (the mouthwash described above used these abilities in hunting down harmful bacteria). They could float and circulate, as antibiotics do, while searching for intruders to neutralize. To escape being engulfed by white blood cells making their own patrols, immune machines could display standard molecules on their surface-molecules the body knows and trusts already—like a fellow police officer wearing a familiar uniform.

When an invader is identified, it can be punctured, letting its contents spill out and ending its effectiveness. If the contents were known to be hazardous by themselves, then the immune machine could hold on to it long enough to dismantle it more completely.

How will these devices know when it's time to depart? If the physician in charge is sure the task will be finished within, say, one day, the devices prescribed could be of a type designed to fall apart after twenty-four hours. If the treatment time needed is variable, the physician could monitor progress and stop action at the appropriate time by sending a specific molecule—aspirin perhaps, or something even safer—as a signal to stop work. The inactivated devices would then be cleared out along with other waste eliminated from the body.

In most parts of the body, the finest blood vessels, capillaries, pass within a few cell diameters of every point. Certain white blood cells can leave these vessels to move among the neighbouring cells. Immune machines and similar devices, being even smaller, could do likewise. In some tissues, this will be easy, in some harder, but with careful design and testing, essentially any point of the body should become accessible for healing repairs.

Merely fighting organisms in the bloodstream would be a major advance, cutting their numbers and inhibiting their spread. Roving medical nanomachines, though, will be able to hunt down invaders throughout the body and eliminate them entirely.

Immunity Systems

Cancers are a prime example. The immune system recognizes and eliminates most potential cancers, but some get by. Physicians can recognize cancer cells by their appearance and by molecular markers, but they cannot always remove them all through surgery, and often cannot find a selective poison. Immune machines, however, will have no difficulty identifying cancer cells, and will ultimately be able to track them down and destroy them wherever they may be growing. Destroying every cancer cell will cure the cancer.

Bacteria, protozoa, worms, and other parasites have even more obvious molecular markers. Once identified, they could be destroyed, ridding the body of the disease they cause. Immune machines thus could deal with tuberculosis, strep throat, leprosy, malaria, amoebic dysentery, sleeping sickness, river blindness, hookworm, flukes, Candida, valley fever, antibiotic-resistant bacteria, and even athlete's foot. All are caused by invading cells or larger organisms (such as worms). Health officials estimate that parasitic diseases, common in the Third World, affect more than one billion people. For many of these diseases, no satisfactory drug treatment exists. All can eventually be eliminated as threats to human health by a sufficiently advanced form of nanomedicine.

Destroying invaders will be helpful, but injuries and structural problems pose other problems. Truly advanced medicine will be able to build up and restructure tissues. Here, medical nanodevices can stimulate and guide the body's own construction and repair mechanisms to restore healthy tissue.

What is healthy tissue? It consists of normal cells in normal patterns in a normal matrix all organized in a normal relationship to the surrounding tissues. Surgeons today (with their huge, crude tools) can fix some problems at the tissue level. A wound disrupts the healthy relationship between two different pieces of tissue, and surgical glues and sutures can partly remedy this problem by holding the tissues in a position that promotes healing. Likewise, coronary artery bypass surgery brings about a more healthy overall configuration of tissues—one that provides working plumbing to supply blood to the heart muscle. Surgeons cut and stitch, but then they must rely on the tissue to heal its wounds as best it can.

Healing establishes healthy relationships on a finer scale. Cells must divide, grow, migrate, and fill gaps. They must reorganize to form properly connected networks of fine blood vessels. And cells must lay down materials to form the structural, intercellular matrix—collagen to provide the proper shape and toughness, or mineral grains to provide rigidity, as in bone. Often, they lay down unwanted scar tissue instead, blocking proper healing. With enough knowledge of how these processes work (and nanoinstruments can help gather that knowledge) and with good enough software to guide the process—a more difficult challenge—medical nanomachines will be able to guide this healing process. The problem here is to guide the motion and behaviour of a mob of active, living cells—a process that can be termed *cell herding*.

Cells respond to a host of signals from their environment: to chemicals in the surrounding fluids, to signal molecules on neighbouring cells, and to mechanical forces applied to them. Cell-herding devices would use these signals to spur cell division where it is needed and to discourage it where it is not. They would nudge cells to encourage them to migrate in appropriate directions, or would simply pick them up, move them along, and deliver them where needed, encouraging them to nestle into a proper relationship with their neighbours. Finally, they would stimulate cells to surround themselves with the proper intercellular-matrix materials. Or like the owner of a small dog who, on a cold day, wraps the beast in a wool jacket—they would directly build the proper surrounding structures for the cell in its new location.

In this way, cooperating teams of cell-herding devices could guide the healing or restructuring of tissues, ensuring that their cells form healthy patterns and a healthy matrix and that those tissues have a healthy relationship to their surroundings. Where necessary, cells could even be adjusted internally, as we will discuss later.

Again, skin provides easy examples and may be a natural place to start in practice. People often want hair where they have bare skin, and bare skin where they have hair. Cell hording machines could move or destroy hair follicle cells to eliminate an unwanted hair, or grow more of the needed cells and arrange them into a working follicle where a hair is desired. By adjusting the size of the follicle and the properties of some of the cells, hairs could be made coarser, or finer, or straighter, or curlier. All these changes would involve no pain, toxic chemicals, or stench. Cell-herding devices could move down into the living layers of skin, removing unwanted cells, stimulating the growth of new cells, narrowing unnaturally prominent blood vessels, insuring good circulation by guiding the growth of any needed normal blood vessels, and moving cells and fibers around so as to eliminate even deep wrinkles.

At the opposite end of the spectrum, cell hording will revolutionize treatment of life-threatening conditions. For example, the most common cause of heart disease is reduced or interrupted supply of blood to the heart muscle. In pumping oxygenated blood to the rest of the body, the heart diverts a portion for its own use though the coronary arteries. When these blood vessels become constricted, we speak of coronary-artery disease. When they are blocked, causing heart muscle tissue to die, we speak of someone "having a coronary," another term for heart attack.

Devices working in the bloodstream could nibble away at atherosclerotic deposits, widening the affected blood vessels. Cell hording devices could restore artery walls and artery linings to health, by ensuring that the right cells and supporting structures are in the right places. This would prevent most heart attacks.

But what if a heart attack has already destroyed muscle tissue, leaving the patient with a scarred, damaged, and poorly functioning heart? Once again, cell-herding devices could accomplish repairs, working their way into the scar tissue and removing it bit by bit, replacing it with fresh muscle fiber. If need be, this new fiber can be grown by applying a series of internal molecular stimuli to selected heart muscle cells to "remind" them of the instructions for growth that they used decades earlier during embryonic development.

Cell-herding capabilities should also be able to deal with the various forms of arthritis. Where this is due to attacks from the body's own immune system, the cells producing the damaging antibodies can be identified and eliminated. Then a cell-herding system would work inside the joint where it would remove diseased tissues, calcified spurs, and so forth, then rework patterns of cells and intercellular material to form a healthy, smoothly working, and pain-free joint. Clearly, learning to repair hearts and learning to repair joints will have some basic technologies in common, but much of the research and development will have to be devoted to specific tissues and specific circumstances. A similar process—but again, specially adapted to the circumstances at hand—could be used to strengthen and reshape bone, correcting osteoporosis.

In dentistry, this sort of process could be used to fill cavities, not with amalgam, but with natural dentin and enamel. Reversing the ravages of periodontal disease will someday be straightforward, with nanomedical devices to clean pockets, join tissues, and guide regrowth. Even missing teeth could be regrown, with enough control over cell behaviour. Moving through tissues without leaving a trail of disruption will require devices able to manipulate and direct the motions of cells, and to repair them. Much remains to be learned—and will be easy to learn with nanoscale tools—but today's knowledge of cells is enough for a start on the problem of how to do surgery on cells.

Cell biology is a booming field, even today. Cells can be made to live and grow in laboratory cultures if they are placed in a liquid with suitable nutrients, oxygen, and the rest. Even with today's crude techniques, much has been learned about how cells respond to different chemicals, to different neighbours, and even to being poked and cut with needles. Conducting a rough sort of surgery on individual cells has been routine for many years in scientific laboratories.

Today, researchers can inject new DNA into cells using a tiny needle; small punctures in a cell membrane automatically reseal. But both these techniques use tools that on a cellular scale are large and clumsy—like doing surgery with an ax or a wrecking ball, instead of a scalpel. Nanoscale tools will enable medical procedures involving delicate surgery on individual cells.

Some viral diseases will respond to treatments that destroy viruses in the nose and throat, or in the bloodstream. The flu and common cold are examples. Many others would be greatly improved by this, but not eliminated. All viruses work by injecting their genes into a cell and taking over its molecular machinery, using it to produce more viruses. This is part of what makes viral illnesses so hard to treat—most of the action is performed by the body's own molecular machines, which can't be interfered with on a wholesale basis. When the immune system deals with a viral illness, it both attacks free virus particles before they enter cells, and attacks infected cells before they can churn out too many more virus particles.

Some viruses, though, insert their genes among the genes of the cell, and lay low. The cell can seem entirely normal to the immune system, for months or years, until the viral genes are triggered into action and begin the infective process anew. This pattern is responsible for the persistence of herpes infections, and for the slow, deadly progress of AIDS.

These viruses can be eliminated by molecular-level cellular surgery. The required devices could be small enough to fit entirely within the cell, if need be. Greg Fahy, who heads the Organ Cryopreservation Project at the American Red Cross's Jerome Holland Transplantation Laboratory, writes, "Calculations imply that molecular sensors, molecular computers, and molecular effectors can be combined into a device small enough to fit easily inside a single cell and powerful enough to repair molecular and structural defects (or to degrade foreign structures such as viruses and bacteria) as rapidly as they accumulate. There is no reason such systems cannot be built and function as designed."

Equally well, a cell surgery device located outside a cell could reach through the membrane with long probes. At the ends of the probes would be tools and sensors along with, perhaps, a small auxiliary computer. These would be able to reach through multiple membranes, unpackage and uncoil DNA, read it, repackage it, and recoil it, "proofreading" the DNA by comparing the sequences in one cell to the sequences of other cells.

On reading the genetic sequence spelling out the message of the AIDS virus, a molecular surgery machine could be programmed to respond like an immune machine, destroying the cell. But it would seem to make more sense simply to cut out the AIDS virus genes themselves, and reconnect the ends as they were before infection. By doing this, and killing any viruses found in the cell, the procedure would restore the cell to health.

Molecular Repairs

Cells are made of billions of molecules, each built by molecular machines. These molecules self-assemble to form larger structures, many in dynamic patterns, perpetually disintegrating and reforming. Cell-surgery devices will be able to make molecules of sorts that may be lacking, while destroying molecules that are damaged or present in excess. They will be able not only to remove viral genes, but to repair chemical and radiation-caused damage to the cell's own genes. Advanced cell surgery devices would be able to repair cells almost regardless of their initial state of damage.

By activating and inactivating a cell's genes, they will be able to stimulate cell division and guide what types of cells are formed. This will be a great aid to cell herding and to healing tissues.

As surgeons today rely on the spontaneous, self-organizing ability of cells and tissues to join and heal the parts they manipulate, so cell-surgery devices will rely on the spontaneous self-organizing capabilities of molecules to join and "heal" the parts they put together. Healing of a surgical wound involves sweeping up dead cells, growing new cells, and a slow and genuinely painful process of tissue reorganization. In contrast, the joining of molecules is almost instantaneous and occurs on a scale far below that of the most sensitive pain receptor. "Healing" will not begin after the repair devices have done their work, as it does in conventional surgery: rather, when they complete their work, the tissue will have been healed.

The ability to herd cells and to perform molecular repairs and cell surgery will open new vistas for medicine. These abilities apply on a small scale, but their effects can be large scale.

In many diseases, the body as a whole suffers from misregulation of the signaling molecules that travel through its fluids. Many are rare: Cushing's disease. Grave's disease, Paget's disease, Addison's disease. Conn's syndrome, Prader-Labhart-Willi syndrome. Others are common: millions of older women suffer from osteoporosis, the weakening of bones that can accompany lowered estrogen levels.

Diabetes kills frequently enough to rank in the top ten causes of death in the United States; the number of individuals known to have it doubles every fifteen years. It is the leading cause of blindness in the United States, with other complications including kidney damage, cataracts, and cardiovascular damage. Today's molecular medicine tries to solve these troubles by supplying missing molecules: diabetics inject additional insulin. While helpful, this doesn't cure the disease or eliminate all symptoms. In an era of molecular surgery, physicians could choose instead to repair the defective organ, so it can regulate its own chemicals again, and to readjust the metabolic properties of other cells in the body to match. This would be a true healing, far better than today's partial fix.

Only now are researchers making progress on another frequent problem of metabolic regulation: obesity. Once this was thought to have one simple cause (consuming excess calories) and one main result (greater roundness than favoured by today's aesthetics), but both assumptions proved wrong. Obesity is a serious medical problem, increasing the risk of diabetes mellitus, osteoarthritis, degenerative diseases of the heart, arteries, and kidneys, and shortening life expectancy. And the supposed cause, simple overeating, has been shown to be incorrect—something dieters had always suspected, as they watched thinner colleagues gorge and yet gain no weight.

The ability to lay in stores of fat was a great benefit to people once upon a time, when food supplies were irregular, nomadism and marauding bands made food storage difficult and risky, and starvation was a common cause of death. Our bodies are still adapted to that world, and regulate fat reserves accordingly. This is why dieting often has perverse effects. The body, when starved, responds by attempting to build up greater reserves of fat at its next opportunity. The main effect of exercise in weight reduction isn't to burn up calories, but to signal the body to adapt itself for efficient mobility.

Obesity therefore seems to be a matter of chemical signals within the body, signals to store fat for famine or to become lean for motion. Nanomedicine will be able to regulate these signals in the bloodstream, and to adjust how individual cells respond to them in the body. The latter would even make possible the elusive "spot reduction program" to reshape the distribution of body fat. Here, as with many potential applications of nanotechnology, the problem may be solved by other means first. Some problems, though, will almost surely require nanomedicine.

New Organs

So far we've seen how medical nanotechnology would be used in the simpler applications outside tissues—such as in the blood—then inside tissues, and finally inside cells. Consider how these abilities will fit together for victims of automobile and motorcycle accidents.

Nanomanufactured medical devices will be of dramatic value to those who have suffered massive trauma. Take the case of a patient with a crushed or severed spinal cord high in the back or in the neck. The latest research gives hope that when such patients are treated promptly after the injury, paralysis may be at least partially avoidable, sometimes. But those whose injuries weren't treated—including virtually all of today's patients—remain paralyzed. While research continues on a variety of techniques for attempting to aid a spontaneous healing process, prospects for reversing this sort of damage using conventional medicine remain bleak.

With the techniques discussed above, it will become possible to remove scar tissue and to guide cell growth so as to produce healthy arrangements of the cells on a microscopic scale. With the right molecular-scale poking and prodding of the cell nucleus, even nerve cells of the sorts found in the brain and spinal cord can be induced to divide. Where nerve cells have been destroyed, there need be no shortage of replacements. These technologies will eventually enable medicine to heal damaged spinal cords, reversing paralysis.

The ability to guide cell growth and division and to direct the organization of tissues will be sufficient to regrow entire organs and limbs, not merely to repair what has been damaged. This will enable medicine to restore physical health despite the most grievous injuries.

If this seems hard to believe, recall that medical advances have shocked the world before now. To those in the past, the idea of cutting people open with knives *painlessly* would have seemed miraculous, but surgical anesthesia is now routine. Likewise with bacterial infections and antibiotics, with the eradication of smallpox, and the vaccine for polio: Each tamed a deadly terror, and each is now half-forgotten history. Our gut sense of what seems likely has little to do with what can and cannot be done by medical technology. It has more to do with our habitual fears, including the fear of vain hopes. Yet what amazes one generation seems obvious and even boring to the next. The first baby born after each breakthrough grows up wondering what all the excitement was about.

Besides, nanoscale medicine won't be a cure-all. Consider a fifty-year-old mentally retarded man, with a mind like a two-year-old's, or a woman with a brain tumor that has spread to the point that her personality has changed: How could they be "healed"? No healing of tissues could replace a missed lifetime of adult experience, nor can it replace lost information from a severely damaged brain. The best physicians could do would be to bring the patients to some physically healthy condition. One can wish for more, but sometimes it won't be possible.

Primary Care

Throughout the centuries, medicine has been constrained to maintain functioning tissues, since once tissues stop functioning, they can't heal themselves. With molecular surgery to carry out the healing directly, medical priorities change drastically—function is no longer absolutely necessary. In fact, a physician able to use molecular surgery would prefer to operate on nonfunctioning, structurally stable tissue than on tissue that has been allowed to continue malfunctioning until its structure was lost.

Brain tumors are an example: They destroy the brain's structure, and with it the patient's skills, memories, and personality. Physicians in the future should be able to immediately interrupt this process, to stop the functioning of the brain to stabilize the patient for treatment.

Techniques available today can stop tissue function while preserving tissue structure. Greg Fahy, in his work on organ preservation at the American Red Cross, is developing a technique for vitrifying animal kidneys—making them into a low-temperature, crystal-free glass—with the goal of maintaining their structure such that, when brought back to room temperature, they can be transplanted. Some kidneys have been cooled to -30°C, warmed back up, and then functioned after transplantation.

A variety of other procedures can also stabilize tissues on a long-term basis. These procedures enable many cells—but not whole tissues—to survive and recover without help; advanced molecular repair and cell surgery will presumably tip the balance, enabling cells, tissues, and organs to recover and heal. When applied to stabilizing a whole patient, such a condition can be called *biostasis*. A patient in biostasis can be kept there indefinitely until the required medical help arrives. So in the future, the question "Can this patient be restored to health?" will be answered "Yes, if the patient's brain is intact, and with it the patient's mind."

Sandra Lee Adamson of the National Space Society has her eyes on distant goals. Some have proposed that travel to the stars would take generations, preventing anyone on Earth from ever making the trip. But she notes that biostasis will "give hope to some fearless adventurers who will risk suspension and subsequent reanimation so they can see the stars for themselves."

Medical nanotechnologies promise to extend healthy life, but if history is any guide, they may also avert sudden massive death. The word plague is rarely heard today, except in relation to AIDS; it calls up visions of the Black Death of the Middle Ages, when one third of Europe died in 1346-50. A virulent influenza struck in 1918, half lost in the news of the First World War: how many of us realize that it killed at least 20 million? People often act as though plagues were gone for good, as if sanitation and antibiotics had vanquished them. But as doctors are forever telling their patients, antibiotics kill bacteria, but are useless for viruses. The flu, the common cold, herpes, and AIDS—none has a really effective treatment, because all are caused by viruses. In some African countries, as much as 10 percent of the population is estimated to be infected with the AIDS-causing HIV virus. Without a cure soon, the steep rise in deaths from AIDS still lies in the future. AIDS stands as a grim reminder that the great plagues of history are not behind us.

New diseases continue to appear today as they have throughout history. Today's population, far larger than that of any previous century, provides a huge, fertile territory for their spread.

Today's transportation systems can spread viruses from continent to continent in a single day. When ships sailed or churned their way across the seas, an infected passenger was likely to show full-blown disease before arrival, permitting quarantine. But few diseases can be guaranteed to show themselves in the hours of a single aircraft flight.

So far as is known, every species of organism, from bacterium to whale, is afflicted with viruses. Animal viruses sometimes "jump the species gap" to infect other animals, or people. Most scientists believe that the ancestors of the AIDS virus could, until recently, infect only certain African monkeys. Then these viruses made the interspecies jump. A similar jump occurred in the 1960s when scientists in West Germany, working with cells from monkeys in Uganda, suddenly fell ill. Dozens were infected, and several died of a disease that caused both blood clots and bleeding, caused by what is now named the Marburg virus. What if the Marburg virus had spread with a sneeze, like influenza or the common cold?

We think of human plagues as a health problem, but when they hit our fellow species, we tend to see them from an environmental perspective. In the late 1980s, over half the harbor-seal population in large parts of the North Sea suddenly died, leading many at first to blame pollution. The cause, though, appears to be a distemper virus that made the jump from dogs. Biologists worry that the virus could infect seal species around the world, since distemper virus can spread by aerosols—that is, by coughing—and seals live in close physical contact. So far its mortality rate has been 60 to 70 percent.

What of AIDS itself: Could it change and give rise to a form able to spread, say, as colds do? Nobel Laureate Howard M. Temin has said, "I think that we can very confidently say that this can't happen." Nobel Laureate Joshua Lederberg, president of Rockefeller University in New York City, replied, "I don't share your confidence about what can and cannot happen." He points out that "there is no reason a great plague could not happen again. We live in evolutionary competition with microbes—bacteria and viruses. There is no guarantee that we will be the survivors."

Bacterial diseases are mostly controllable today. Sanitation limits the ways in which plague can spread. These measures are just good enough to lull us into imagining the problem is solved.

Viruses are common, viruses mutate; some spread through the air, and some are deadly. Plagues show that fast-spreading diseases can be deadly, and effective antiviral drugs are still rare.

The only really effective treatments for viral diseases are preventive, not curative. They work either by preventing exposure, or by exposing the body beforehand to dead or harmless or fragmentary forms of the virus, to prepare the immune system for future exposure. As the long struggle for an AIDS vaccine shows, one cannot count on modern medicine to identify a new virus and produce an effective vaccine within a single month or year or even a single decade. But influenza epidemics spread fast, and Marburg II or AIDS II or something entirely new and deadly may do the same.

The deaths from the next great plague could have begun in a village last week, or could begin next year, or a year before we learn to deal with new viral illnesses promptly and effectively. With luck, the plague will wait until a year after.

Immune machines could be set to kill a new virus as soon as it is identified. The instruments nanotechnology brings will make viral identification easy. Some day, the means will be in place to defend human life against viral catastrophe.

From eliminating viruses to repairing individual cells, improving our control of the molecular world will improve health care. Immune machines working in the bloodstream seem about as complex as some engineering projects human beings have already completed—projects like large satellites. Other medical nanotechnologies seem to be of a higher order of complexity.

Somewhere in the progression from relatively simple immune devices to molecular surgery, we've crossed the fuzzy line between systems that teams of clever biomedical engineers could design in a reasonable length of time and ones that might take decades or prove impossibly complex. Designing a nanomachine capable of entering a cell, reading its DNA, finding and removing a deadly viral DNA sequence, and then restoring the cell to normal would be a monumental job. Such tasks are advanced applications of nanotechnology, far beyond mere computers, manufacturing equipment, and half-witted "smart materials."

To succeed within a reasonable number of years, we may need to automate much of the engineering process, including software engineering. Today's best expert systems are nowhere near sophisticated enough. The software must be able to apply physical principles, engineering rules, and fast computation to generate and test new designs. Call it *automated engineering*.

Automated engineering will prove useful in advanced nanomedicine because of the sheer number of small problems to be solved. The human body contains hundreds of kinds of cells forming a huge number of tissues and organs. Taken as a whole (and ignoring the immune system), the body contains hundreds of thousands of different kinds of molecules. Performing complex molecular repairs on a damaged cell might require solving millions of separate, repetitive problems. The molecular machinery in cell surgery devices will need to be controlled by complex software, and it would be best to be able to delegate the task of writing that software to an automated system. Until then, or until a lot of more conventional design work gets done, nanomedicine will have to focus on simpler problems.

Where does aging fit in the spectrum of difficulty? The deterioration that comes with aging is increasingly recognized as a form of disease, one that weakens the body and makes it susceptible to a host of other diseases. Aging, in this view, is as natural as smallpox and bubonic plague, and more surely fatal. Unlike bubonic plague, however, aging results from internal malfunctions in the molecular machinery of the body, and a medical condition with so many different symptoms could be complex.

Surprisingly, substantial progress is being made with present techniques, without even a rudimentary ability to perform cell surgery in a medical context. Some researchers believe that aging is primarily the result of a fairly small number of regulatory processes, and many of these have already been shown to be alterable. If so, aging may be tackled successfully before even simple cell repair is available. But the human aging process is not well enough understood to enable a confident projection of this; for example, the number of regulatory processes is not yet known. A thorough solution may well require advanced nanotechnology-based medicine, but a thorough solution seems possible. The result would not be immortality, just much longer, healthier lives for those who want them.

A challenging problem related to medicine (and to biostasis) is that of species restoration. Today, researchers are carefully preserving samples from species now becoming extinct. In some cases, all they have are tissue samples. For other species, they've been able to save germ cells in the hope that they will be able to implant fertilized eggs into related species and thus bring the (nearly?) extinct species back.

Each cell typically contains the organism's complete genetic information, but what can be done with this? Many researchers today collect samples for preservation thinking only of the implantation scenario: one that they know has already been made to work. Other researchers are taking a broader view: the Center for Genetic Resources and Heritage at the University of Queensland is a leader in the effort. Daryl Edmondson, coordinator of the gene library, explains that the center is unique because it will "actively collect data. Most other libraries simply collate their own collections." Director John Mattick describes it as a "genetic Louvre" and points out that if genes from today's endangered species aren't preserved, "subsequent generations will see we had the technology to keep [DNA] software and will ask why we didn't do it." With this information and the sorts of molecular repair and cell-surgery capabilities we have discussed, lost species can someday be returned to active life again as habitats are restored.

Some Uses of Nanotechnology in Medicine

This final Chapter describes a miscellany of important technical capabilities that may prove useful in some or all medical nanodevices, in various scenarios or theaters of operation. Any one of these subjects deserves an entire chapter to itself, but unfortunately there is only space in this introductory text for a brief survey of each area. The most important of these topics is computation, including nanomechanical, nanoelectronic, and biological computing, as well as nanoscale data storage technologies. However, the fields of organic and fullerene nanoelectronics, biocomputing, and quantum computing are advancing so fast that whatever is written here will quickly become obsolete. Thus our coverage in this Volume is limited to a broad overview. Interested readers are strongly advised to consult the current literature for the latest results.

Other important basic capabilities covered in this Chapter include timers and nanoclocks with long-term nanosecond stability; high-pressure materials storage ; defensive cytocidal and antiviral weaponry to allow the efficient destruction of pathogenic intruders and unrepairable tumor cells; and the effects of very hot or very cold temperatures on nanorobot materials and operations.

Nanochronometry

Nanorobots will use clocks in many applications. Computer gating, navigation, and high-speed sensor applications may require repeatable timing in the 1-1000 nanosec range, or 1-1000 microsec for lower-speed chemical and chemotactic sensing, although long-term clock stability is not especially critical for computing. Neuron-mediated signals and muscle motions are monitored and controlled on a 1-1000 millisec timescale, while conscious human action and most human biorhythms occur in the 1-10^5 sec range.

Human Chronobiology

The human body incorporates numerous biological clocks. The best-known and most-studied example is the daily (24-hour) endogenous circadian oscillator. This internal clock is normally reset by natural sunlight, which is much brighter than indoor lighting. The clock, in turn, sets the cadence for most of the other 24-hour body rhythms — for example, sleep/wakefulness cycles (e.g., melatonin), urine production, body temperature, blood cortisol and ACTH cycles, and the diurnal rhythm of mitosis in epidermal epithelium (e.g., greatest during sleep or inactivity, least during wakefulness or activity). The circadian clock stability (variation in free-running period) is ~1 % in *Drosophila* and ~0.2% in humans.

Mammalian circadian rhythms are regulated by a master pacemaker within the suprachiasmatic nuclei (SCN) of the hypothalamus (just above the point at which the optic nerves from each eye cross in midbrain). In humans, the SCN is comprised of ~10,000 special cells that send out electrochemical signals in a 24-hour timing pattern. Entrainment of the clock to light-dark cycles is mediated by photoreceptors in the retina, with light information conveyed directly from ganglion cells of the retina to the SCN via the retinohypothalamic tract. Mammals and other vertebrates have another independent circadian clock in each retina, possibly driven by the retinal cryptochromes CRY1 (also made in the SCN) and CRY2, producing rhythms in local physiology such as the diurnal renewal cycles of rods and cones. CRY1 is also abundant in skin tissues. Clock-resetting photoreceptors have been found in the human popliteal region (behind the knees), and such photoreceptors may exist in many different tissues throughout the human body.

The current model for a core circadian oscillator comprises, in part, a transcription/translation-based negative feedback loop in which clock genes are rhythmically expressed, giving rise to cycling levels of clock RNAs and proteins (negative elements). The proteins then feed back, after a lag, to depress the level of their own transcripts, perhaps by interfering with positive elements that increase transcription of the clock genes. In 1998 this genetic network was still in the discovery process. One ~100,000-base-pair gene, aptly named "clock," was known to produce an 855-residue protein (mCLOCK in mice, 115.7 kD) that serves as a major regulator of the ~10 other, still unidentified, genes thought to affect circadian rhythm. (One portion of mCLOCK is the same as in a related 1023-residue clock protein, dCLOCK, found in *Drosophila* fruit flies). Circadian rhythms will almost certainly be made up of many interconnected feedback loops that interact with the CLOCK-related core pathway. It is believer that this interconnected ensemble will ultimately determine all of the classical circadian properties—period length, temperature compensation, and resetting by light or temperature. In 1998, most chronobiologists believed that many of these outer loops would be organism specific and that only the core loop would be more universal.

In addition to the 24-hour circadian oscillator, many other biological clocks have been identified in humans. For example, the 28-day menstrual cycle is timed by two ovarian clocks, each delimiting a period of 14 days. One of these two clocks is the Graafian follicle and its production of estradiol, while the other clock is the corpus luteum and its secretion of progesterone. Both are obligatorily dependent upon the proper functioning of a third clock located in the arcuate region of the hypothalamus, known as the GnRH pulse generator, which ensures the rhythmic production and release of the gonadotropic hormones into the peripheral circulation with a cyclical period of ~1 hour (± 25% stability), coincident with a 1-hour cycle of hypothalamic electrical activity (± 12% stability). A 90-minute clock paces the development of the somites, blocks of tissue that form in regular arrays along the spinal cord of vertebrate embryos. One regulatory gene, named "chairy," undergoes repeated 90-minute activity cycles and is known not to require protein synthesis; gene expression follows the repeating pattern even when protein synthesis is biochemically blocked. A much longer half-weekly (circasemiseptan) pattern has been observed in mitotic activity in cancer patients.

Other biological clocks abound. The stomach produces rhythmic contractions of ~3/minute during digestion. Rhythmic segmentation contractions of the bowel at ~0.2 Hz are driven by a thin layer of specialized pacemaker cells called the intestinal cells of Cajal, which demonstrate electrical oscillations at about the same rate as the contractions. Cardiac pacemaker cells in the sinoatrial node emit a cyclical pulse that triggers a heartbeat at a ~1 Hz basal rate. The respiratory centers in the medulla oblongata produce an autonomic contraction cycle of the diaphragm at ~0.3 Hz. The human eye produces many spontaneous rhythmic dilations and constrictions, including high-frequency ocular microtremors present in all people (mean frequency ~84-88 Hz), accommodation microfluctuations (~1-2 Hz) with the pupil varying <300 microns in diameter, and low-frequency pupillary oscillations (e.g., ~0.2 Hz "hippus"). Electrical rhythms generated by the brain that are detectable by an electroencephalograph include alpha waves predominantly in the occipital region at 8-13 Hz, beta waves in the frontal and central areas at 18-30 Hz, delta waves at <4 Hz during deep sleep or abnormal function, and theta waves in the temporal and parietal areas at 4-7 Hz.

Individual cells display a number of oscillatory biochemical pathways as well, including calcium ion oscillations, cyclic AMP signalling, and various other cell cycles. For instance, the glycolytic enzyme oscillator, controlled primarily by phosphofructokinase, may mediate various rhythmic physiological behaviours such as slow waves of contraction in smooth muscle, electrical activity in neurons, and insulin release from b islet cells of the pancreas. In one experiment, this oscillator mediated a 55 mV voltage change in guinea pig cardiomyocytes at 0.010 Hz (±10% stability) in an almost perfectly sinusoidal pattern, although a few cells displayed oscillations with irregular phase, amplitude, or both. In another experiment, fluorescently-labeled puffs of tumor-cell cytoplasm were observed being released through plasma membrane ruptures, regularly at ~0.05 Hz during a neutrophil-mediated cytolytic attack, matching the nicotinamide-adenine dinucleotide phosphate and superoxide release periods for neutrophils. By 1998 there was a growing belief that many, if not most, of the cells in an animal may possess individual biological clocks; fruit flies were already known to have clocks distributed throughout their wings, legs, and abdomen.

The human brain also possesses an interval timer that allows a person to gauge the passage of seconds or minutes to a mean accuracy of ±15%. The interval timer functions like a stopwatch and resides in the basal ganglia, a region of the brain that coordinates voluntary muscle movements. A population of neurons in the substantia nigra releases regular pulses of dopamine into an accumulator, called the caudate-putamen, a major part of the basal ganglia. These pulses are then read via neural pathways to the cerebral cortex which are known as striatocortical loops. Other internal "alarm clocks" that reliably allow waking from sleep at desired times have been described. In vivo nanorobots should be able to eavesdrop on most of these active chronobiological channels, gaining complete knowledge of die oscillator frequencies and phase settings of virtually all of the human body's biological clock systems. For example, blood plasma cortisol typically peaks in the morning (e.g., 6-23 × 10^{-8} gm/cm^3 at 8 AM), then declines throughout the day (e.g., 3-15 × 10^{-8} gm/cm^3 at 4 PM, then ~50% of 8 AM value at 10 PM). Melatonin (also detectable in saliva) cycles in the opposite direction, normally peaking at 6-7 × 10^{-11} gm/cm^3 at night and then falling to 1.4×10^{-11} gm/cm^3 during the day. Longer cycles are easily tracked too. For instance, blood viscosity in healthy young women shows the rhythmic variation of the menstrual cycle, probably due to changes in serum fibrinogen and globulin levels which can be measured more directly. These nanorobot-accessible readings may be used either to measure "body time" or to control it, and in some cases can recalibrate relevant onboard nano-chronometers.

Environmental time cues may also be detected by in vivo nanorobots (e.g., temporal macrosensing), such as the level of illumination penetrating the epidermis (sometimes permitting inference of day or night) or the all-pervasive 60/50 Hz (U.S./Europe) electromagnetic hum generated by alternating current electrical supply systems in common use throughout the world, a crude frequency standard whose voltage can sometimes display time-of-day load-related fluctuations.

Artificial Oscillators

Nanorobot clocks must perform at least two critical functions: (1) Interval timing (measuring elapsed time between two or more sequential events) and chronometry (maintaining an onboard calendrical "standard time" in good calibration with an external time standard such as Coordinated Universal Time). Complete clock design, including clutches, feedback mechanisms and governors, dampers and filters, triggers and strikers, amplifiers and mixers, phase detectors and counters, regenerative frequency dividers and frequency multipliers, and anti-vibration housings, is very complex and quite beyond the scope of this text. However, both timing and chronometry require a high-frequency oscillator of known and stable frequency, whose oscillations can then be counted for clocking purposes. If frequency multipliers can be built, then the primary oscillator need not be high frequency; the requirement for frequency stability is more fundamental. Timing accuracies of 1-1000 nanosec are typically needed for sensor, navigation, and computational applications, with operating frequencies up to $n_{osc} \sim 1$ GHz and durations ranging from 1 microsec up to 10^5 sec ($\sim$1 day) between timed events or between clock recalibrations. This Section presents a brief survey of several useful nanoscale oscillator systems.

Mechanochemical and photochemical oscillators

A chemical oscillator involves transitions between two well-defined chemical, energetic, or conformational states. For example, repetitive room-temperature protein folding such as a-helical coiling may occur in $\sim 10^{-6}$ sec (e.g., $n_{osc} \sim 1$ MHz). Turnover number (molecules of substrate converted to product per active site per unit time) for catalase, one of the fastest known enzymes, is $k_{cat} = n_{OSC} = 40$ MHz. Reversible gas phase reactions may occur at up to 5 GHz. However, chemical-based clocks are expected to display poor frequency stability. Most chemical reactions display $\sim e^{-1/T}$ rate dependence, giving a frequency stability of Dn/n ~ 1% for a ~1 K temperature (T) variation in the oscillator near 310 K. Chemically-driven mechanical oscillators such as biological cilia display a similar temperature dependency and frequency stability. Time measurements that depend on the rates of thermally activated bulk reactions also suffer Poisson noise in the reaction rate from the statistics of interaction of many random independent molecules.

Some molecular transitions are even faster. Molecular dynamics simulations of carbonmonoxy-myoglobin show periodic openings and closings of internal voids on a time scale of ~100 picosec, giving n_{OSC} ~10 GHz. The torsional isomerization of the retinal chromophore in rhodopsin from the 11-cis to the 11-trans structure, a ~0.5 nm physical motion detectable by displacement nanosensors, occurs within ~0.2 picosec of the absorption of the 500 nm photon (~420 zJ/molecule), potentially allowing event timing at n_{OSC} ~ 5 THz. The frequency of decomposition of activated complexes in classical reaction rate theory is given by Eyringas n_{active} ~ kT/h ~ 6 THz, approximately the characteristic timescale for molecular vibrations. The frequency stability of these transitions is unknown.

Mechanical oscillators

The most accurate macroscale mechanical clock was the Shortt pendulum clock, the primary timekeeper at the Greenwich and U.S. Naval observatories until the late 1940s. The Shortt clock consisted of a pendulum swinging freely in near-vacuum at constant temperature, driven by another identical clock slaved to the first; frequency stability was Dn/n ~ 4×10^{-8}.

In the nanoscale realm, Drexler proposes a n_{OSC} ~ 1 GHz mechanical clock intended to drive a nanomechanical computer CPU. In this clock, a DC electrostatic motor turns a crankshaft that converts rotary motion into sinusoidally oscillating linear motions of a drive rod. A cam surface on the drive rod forces a follower rod into up or down positions, generating regular clock pulses with intervals determined by the position of the follower with respect to the mean position of the ramp on the cam surface. Clock stability is limited by the stability of the crankshaft rotation rate. The proposed DC electrostatic motor has radius R_{motor} = 195 nm, rim speed v_{rim} =

1000 m/sec, and output power P_{motor} = 1.1 microwatt, giving a developed torque of t_{motor} ~ 2 × 10^{-16} N-m.

Acoustic oscillators

Drexler describes a diamondoid acoustic transmission line in which a ~2 nN force pulse is initiated at one end, travels to the far end at v_{sound} ~ 17,300 m/sec, then is received by a mechanical displacement probe, possibly with significant energy recovery. Such lines may also be used to generate precisely delayed acoustic signal sets suitable for clocking applications.

Consider a "starter" pulse applied to a short feeder line that symmetrically bifurcates into a set of n_{lines} = 10 lines each of length l_n = (1730 n) nm (n = 1, 2, ..., n_{line}), requiring a total length L_{line} = (1730) n_{line} (n_{line} + 1)/2 = 95,150 nm of transmission line to obtain a set of lines having all n_{line} time delays. The bifurcated pulses arrive at the end of each line in 0.1 nanosec (n = 1), 0.2 nanosec (n = 2), ..., 1.0 nanosec (n = 10); any of these pulses may be drawn off and used for diverse clocking purposes, or may be fed back into the initiation mechanism and used either to trigger the next starter pulse or to achieve more robust error correction. If acoustic lines have a cross-sectional area of ~30 nm, then the total volume of all ten lines is ~3,000,000 nm, which may be coiled into ~0.3% of the volume of a I micron nanorobot. Such diamondoid acoustic power transmission lines are essentially lossless. If a complete set of lines containing all n_{line} delays is not required, any desired single delay time may be obtained by connecting up to n_{line} segments end to end, or by bouncing a pulse off of the ends of a single line up to n_{line} times.

Aside from the many potential frequency instabilities inherent in pulse detection, signal re-initiation, and acoustic interferometry mechanisms, two fundamental sources of frequency instability include:

1. changes in the velocity of sound due to thermal variations, and;
2. changes in acoustic path length due to elastic longitudinal displacements of thermally excited transmission rods.

Quartz resonators

Piezoelectric quartz crystals, when cut in a predetermined manner and mounted so that two opposing faces have electrical contacts, will compress or expand when an oscillating electric field is applied across the faces. The amplitude of crystal vibration falls off very rapidly as the applied electric field frequency deviates from the resonant frequency of the crystal. (Cyclical mechanical compressions at the resonant frequency also induce oscillating electric fields of maximum amplitude.)

In 1998, most high-precision timing devices were based on quartz crystal resonators, with ~2 billion resonators manufactured annually worldwide for oscillator, clock, and filter applications. Commercially-available off-the-shelf GHz devices (e.g.. Micro Networks' M101 resonator) typically displayed ~10 ppm stability, or Dn/n ~ 10^{-5}, readily improved to ~10^{-7} in temperature-controlled ovens.

The resonant frequencies of the highest-quality, lowest-noise quartz macroresonators can be measured with a precision of 14 significant figures, e.g:, the noise is Dn/n ~ 3 × 10^{-14} at the optimum measurement times. However, each variable in is temperature dependent. For instance, the thermal expansion of quartz (affecting density) and the temperature coefficients of E_{quartz} (ranging from positive to negative values) are highly dependent on the angles of cut of the plate relative to the crystallographic axes. Quartz resonator frequency may vary monotonically with temperature at 10^{-5}-10^{-4} K^{-1}; the noise floor (i.e., the Allan deviation floor) for quartz microresonators has been estimated as Dn/n ~ (1.2 × 10^{-19} Hz^{-1}) n_{OSC} ~ 10^{-10} for n_{OSC} ~ 1 GHz. Atomistic simulation of submicron quartz oscillators by Broughton suggests that such oscillators must have at least ~10^6 atoms to display bulk elastic constants, and that differences from continuum mechanical frequency predictions are observable for 17-nm (or smaller) devices, and that nanoscale devices with even a single defect may exhibit dramatic anharmonicity.

Atomic frequency standards

The most accurate oscillators are the Atomic Frequency Standards (AFS) used in "atomic clocks." In these "clocks," an atom flips between two slightly different configurations—one in which the electron spin and the nuclear spin point in the same direction, and another configuration in which the two spins point in opposite directions. In its original form, a beam of cesium (Cs^{133}) atoms is emitted from an oven and passes through an evacuated (~10^{-11} atm) chamber, where the atoms are focused by one fixed magnet, then defocused by a second fixed magnet. Between the two fixed magnets, the beam passes through a microwave field. When the frequency of this oscillating field exactly matches a natural atomic resonance ground-state hyperfine transition frequency of Cs (~9.192,631,770 GHz), the spin energy state can switch polarity, allowing "flipped" atoms to focus, rather than defocus, at the second fixed magnet. Atoms arriving at the focus are ionized by a hot-wire ionizer target, then directed onto an electron multiplier by a mass spectrometer to be counted. The microwave frequency is adjusted until the electron multiplier output current is maximized, constituting the measurement of the atoms' resonance frequency.

Early cesium AFSs were used to recalibrate a quartz oscillator about once a day, achieving $Dn/n \sim 3 \times 10^{-11}$. In 1998, the best non-cryogenic laboratory Cs or Rb (rubidium) AFS had $Dn/n \sim 2 \times 10^{-14}$, with a temperature stability of $\sim 10^{-13}$ K^{-1} between 263-313 K and a magnetic stability of $\sim 10^{-12}$/gauss. The smallest non-cryogenic Rb AFS was a space-qualified system with mass of 1.3 kg and power consumption of 11 watts ($\sim 10^{4}$ watts/m^{3}), achieving $Dn/n < \sim 5 \times 10^{-13}$. Another system that achieved $Dn/n < \sim 5 \times 10^{-14}$ had mass ~5.5 kg and drew 39 watts. Laser-cooled low-temperature clocks using Bose-Einstein condensates were expected eventually to demonstrate $Dn/n \sim 10^{-18}$.

In 1998, the miniaturization of rubidium AFS for space applications was being actively studied, along with newer approaches such as optically-pumped cesium AFS using solid state diode lasers (thus eliminating the bulky magnets), diode laser-pumped rubidium AFS in which the Rb discharge lamp is replaced with a ~100% efficient diode laser tuned to the correct transition frequency, and Hg^{+} "optical clocks." Optical pumping methods using diode lasers defined the NIST-7 atomic-beam standard of $Dn/n < \sim 5 \times 10^{-15}$ starting in 1993.

A detailed analysis of micron-size atomic clocks is beyond the scope of this book. The possibility cannot be ruled out, but in 1998 the feasibility was unknown. From the minimum sensor capable of detecting a spin transition requires $N_{min} \sim 7 \times 10^{6}$ Cs atoms (~0.001 $micron^{3}$ of Cs), taking transition frequency n_L = 9.192,631,770 GHz and electron spin angular momentum $L_{electron} = L_{proton}$. However, many problems must be overcome including interactions involving spontaneous decay, measurement-induced transitions, phase changes due to gas molecule collisions, spatial confinement effects, the use of symmetries and cancellation of couplings. J. Soreff suggests consideration of an oscillator structure in which a phosphorus atom is covalently bound to a tetrahedral support of carbyne rods extending to the ends of an evacuated chamber. The fifth valence electron on the phosphorus atom should have a hyperfine interaction with the P^{31} nucleus, but it is presently unknown how tightly coupled the thermal vibrations in the rods are to the hyperfine state transition, since many coupling modes may vanish by symmetry.

Nanorobot Synchronization

Ideally, the clocks of all nanorobots present in a patient's body will be synchronized to some universal time. This time setting may be defined by the patient, by the physician, or by reference to some external standard. As part of a chronometry-critical nanorobot installation procedure prior to infusion, an initializing signal can be transmitted throughout the infusate volume, allowing each individual nanorobot present therein to receive the signal and set its clock.

Acoustic synchronization pulses are appropriate if ~microsec precision will suffice. Sound waves passing at v_{sound} = 1500 m/sec through an L_{vial} = 1.5 cm wide container of well-stirred aqueous-suspended nanorobots restricts synchronization error to $Dt_{error} \sim L_{vial}/v_{sound}$ = 10 microsec.

For the highest pre-infusion synchronization accuracy, optical or rf pulses will be employed. Electromagnetic energy travels at c ~ 3 × 10^8 m/sec across the container volume until intercepted by a nanorobot optical sensor, giving $Dt_{error} \sim L_{vial}/c$ = 0.05 nsec. Allowing 100 green photons (360 zJ/photon at 550 nm) per nanorobot at a nanorobot number density of 10^{12} cm^{-3}, the flash injects ~40 J/m^3 of energy into the container, raising the water temperature by ~10 microkelvins after this energy is fully thermalized. A 1-nsec flash produces a peak intensity of ~4 × 10^{10} watts/m, comparable to monochromatic optical tweezers which are tolerated by biological macromolecules. Some nanorobots inevitably will miss the signal and may later synchronize, prior to infusion, via close or direct physical contact with others who received the signal, or else multiple synchronization pulses can be used.

Onboard nanorobot clocks can also be synchronized post-infusion. For example, a pressure cuff inflated around the patient's arm can administer acoustic synchronization pulses to ~microsec precision as bloodborne nanorobots pass through the blood vessels within; after signaling for several mean blood circulation times (e.g., a few minutes), most bloodborne nanorobots should be properly synchronized. Whole-body acoustic transmissions at 0.1-1 MHz may be generated by operating tables, vibrating chairpads, wristwatch transmitters and the like. Such methods may produce synchronization errors of ~100 microsec over 15 cm anteroposterior path lengths, reducible to ~1 microsec error using N_{obs} = 10,000 repetitions of the calibration signal. For higher precision, the physician transmits ~MHz radio wave pulses into the body; such pulses may be detected by appropriate onboard receivers. Such waves have only ~75% attenuation over 15-cm anteroposterior path lengths, producing 0.5 nsec synchronization accuracy. Optical photons applied to the skin surface are subject to scattering, producing signal pulse broadening which increases synchronization error, and to absorption, which makes signal reception difficult or impossible beyond tissue depths of a few centimeters.

Another approach is to inject a relatively small number of chronocytes—mobile communicytes modified to include an onboard nanoclock of high precision (e.g., a portable frequency standard) and the ability to transmit clocking synchronization signals to calibrate neighbouring nanorobots. These signals may be transmitted acoustically at close proximity to the nodes of a mobile communications network, subsequently allowing all nanorobots within 100 microns of a calibrated node to synchronize to within $< \sim 67$ nsec. Averaging algorithms can be employed to synchronize uncalibrated nodes in the system that were not recently visited by a chronocyte. Bloodborne chronocytes will normally be carried from the capillary vasculature to within ~ 100 microns of most tissue locations, thus allowing $< \sim 67$ nsec single-pulse acoustic synchronization even in the absence of an installed communications network.

Dedicated Chronometer Organs

Just as the hypothalamus is the physical locus of the natural circadian clock, it may be convenient to establish artificial chronometer organs inside the human body for various purposes. Dedicated nano-organs have been described previously in connection with power, communication, and navigational systems.

Chronometer organs, which would likely be millimeter-scale or smaller, could be used as endogenous clocks to regulate the precise administration of time-release substances or devices. Such organs could serve as highly accurate timers to improve the natural human interval timing sense from $\pm 15\%$ accuracy in biological systems to parts per million accuracy, or better, using artificial systems. If linked to the patient or user through various out-messaging channels, chronometer organs can provide a continuously-available, consciously-accessible unfailing "time sense" of extraordinary precision for time of day, calendar date, and other time-related information.

Dedicated chronometer organs can be used to synchronize chronocytes or mobile communication networks at the nodes. For example, chronocytes with lower-stability crystal oscillators could be repeatedly recalibrated as they passed near larger embedded chronometer organs during each blood circulation cycle (e.g., once per minute). Such chronometer organs might incorporate higher-stability onboard atomic frequency standards. Alternatively, chronometer organs could serve as transdermal data ports through which timing synchronization signals can be injected into fiber-based in vivo communication networks from external timing sources, similar to the autoclock pulse (containing time and date information) that is added to most contemporary television broadcast signals. Hard connectors, centimeter-scale rf wireless antennas, and other types of links are possible.

In 1998, a desk clock could be purchased for $50 that automatically recalibrated itself by picking up WWVB radio signals anywhere in North America, several times a nighty using a few centimeters of rod antenna. WWV's carrier frequency is regulated to Dn/n ~ 5 × 10^{-12}, ultimately providing a time synchronicity at the receiver of ~100 microsec/day. With a slightly larger in vivo antenna, satellite GPS signals may be received, providing < ~20 nanosec time error. If buildings and vehicles are rewired to permit dissemination of time synchronization pulses and other useful information via continuous rf or IR channels, chronometer organs could employ smaller in vivo antennas and still achieve very precise results.

Although it has been informally speculated that nanorobots could use biological nerve fibers as rf antennas to receive electromagnetic time recalibration signals from outside the human body, such as from WWV, this concept is probably unworkable for several reasons. First, the axoplasm is only ~10^{-7} as electrically conductive as a metal wire of equivalent size, because axoplasmic charge carrier density and mobility are much lower than for electrons in a wire.

Second, passively conducted axonal currents are quickly attenuated by leakage through ion channels in the membrane; which is a very poor insulator. From cable theory applied to neurons, the length constant 1_n is defined as the distance over which an applied potential depolarizes to l/e (~37%) of its maximum value; $1_n = (d_{neuron}\ r_{mem} / 4\ r_{axo})^{1/2} \sim (0.04\ d_{neuron})^{1/2}$ where d_{neuron} is axonal diameter in meters, r_{mem} is the specific membrane resistance (~0.2 ohm-m in human neurons) and r_{axo} is the specific resistance of the axoplasm (~1.25 ohm-m in human neurons). For a typical human axon with d_{neuron} ~ 1 micron, then 1_n ~ 200 microns (~internodal distance between nodes of Ranvier) and an external signal is >99% attenuated in ~1 mm of longitudinal travel.

Third, short pulses (e.g., rapidly-oscillating rf signals) are severely distorted and attenuated by the electrical capacitance of the cell membrane. The capacitive time constant in human nerves and muscle cells ranges from 120 millisec, thus limiting any possible electromagnetic reception to frequencies under 50-200 Hz, although pulsed microwaves directed through the brain can induce auditory effects in animals and humans.

Nanocomputers

Many important medical nanorobotic tasks will require computation during the acquisition and processing of sensor data, the control of tools, manipulators, and motility systems, navigation and communication, and during the coordination of collective activities with neighbouring nanorobots. Ex vivo computation has few theoretical limits, but computation by in vivo nanorobots will be subject to a number of constraints such as physical size, power consumption, onboard memory and processing speed.

The memory required onboard a medical nanorobot will be strongly mission dependent. Recognizing and manipulating molecules is fundamental. A very simple mission might demand only the identification or handling of perhaps ~10 different molecules. For example, basic respiratory gas transport nanorobots such as respirocytes may require the operation of fixed-shape receptors that bind simple molecules such as O_2, CO_2, H_2O, and glucose. Identifiers for such receptors may need only a few bits, hence this memory requirement should be negligible. A toxin removal device similarly may require keeping track of only a few types of fixed-shape receptors. On the other hand, a survey or assay mission might need to recognize N = 100-1000 distinct proteins. A spherical 1-micron nanorobot can have $>10^4$ fixed-shape receptors on its surface; if these will suffice, then we require $N \log_2 (N) \sim 10^4$ bits to identify each of N = 1000 different receptor types. This seems more efficient than using more advanced reconfigurable receptors, which might need $>10^4$ bits per receptor-pattern to specify each binding site geometry to the necessary atomic-scale resolution, thus imposing a total memory requirement of $>10^7$ bits for an onboard library of N = 1000 different receptor types. Consequently, simple missions involving basic process control with limited motility may require no more than $\sim 10^5$-10^6 bits of memory, comparable to an old Apple II computer (including RAM plus floppy disk drive). At the other extreme, a complex cell repair mission might require the onboard storage of a substantial fraction of the patient's genetic code, representing $\sim 10^9$ bits of memory including perhaps $\sim 0.2 \times 10^9$ bits of linear sequence data for all 100,000 protein types found in the human body, again assuming 300 amino acids per protein. (Most amino acids are folded into the protein's interior and are not readily accessible to surface probing unless the protein is unfolded, which usually is not desirable or convenient. On the other hand, binding sites for large molecules should be physically easier to construct than small-molecule receptors. An onboard memory of 10^9-10^{10} bits would be in the same range as the 1985 Cray-2 (2×10^{10} bits) or the 1989 Cray-3 (6×10^8 bits) supercomputers.

Computational speed will also be strongly mission dependent. However, extremely simple process control systems in basic factory settings may only require speeds as slow as 10^4 bit/sec. Individual natural biocomputational devices (as opposed to multiple such biodevices operating in parallel) generally do not exceed this speed. Examples include mRNA translation during protein manufacture at ~15 Hz (~75 bits/sec assuming 5 bits/protein); transcription from DNA by RNA polymerase at ~40 Hz (~80 bits/sec at 2 bits/nucleotide); DNA replication at ~800 Hz (~1600 bits/sec at 2 bits/nucleotide); typically 5-100 Hz (bits/sec) for neural electrical discharges; ~1000 Hz (bits/sec) for excitory cholinergic synapses, and gated ion channels at ~10^4 Hz (bits/sec). At the other extreme, a processing speed of 10^9 bits/sec allows a ~10^9 bit genomic information store to be processed in ~1 sec, the small-molecule diffusion time across an average 20-micron wide cell. In 1998, personal desktop computers capable of ~10^9-10^{10} bits/sec (~10^8 operations/sec) were commonly available.

This Section describes possible nanomechanical and nanoelectronic computers, biocomputers, and briefly examines the ultimate limits to computation including reversible and quantum computing. Computer architectural issues are not explicitly addressed here. (For example, advanced architectures, especially large CPU systems, may employ distributed clocks and a distributed computing network.)

Nanomechanical Computers

Electronic computers were evolutionarily preceded by purely mechanical computational devices, starting with the venerable abacus (hand-operated movable beads on rods) in ~3000 BC, the Pascaline (the first hands-on algorithm-executing machine) in 1642, and the Difference Engine (the first hands-off algorithm-executing machine) designed in 1821 by Charles Babbage. A 2000-part working subsection of the brass-geared Engine was demonstrated in 1832; an entire working Difference Engine was reconstructed by historians in 1991, proving that Babbage's design was sound. In the 1840s, Thomas Fowler built and exhibited a calculating device using sliding rods made of wood instead of metal. Whereas Babbage's engines used the familiar 0-9 decimal system with each number represented by a discrete position of a rotating gear wheel. Fowler's machine was more fully digital, using as its active element not rotating wheels but sliding "trinary" rods which could occupy only one of three positions at any time, the first known example of "rod logic." (By reducing the number of distinct physical states, parts could be made less precisely.)

By 1834, Babbage had also conceived detailed plans for his Analytical Engine, intended as a general-purpose programmable computing machine but based entirely on 19th century mechanical technology. The Analytical Engine was to have a random-access memory consisting of 1000 words of 50 decimal digits each (~175,000 bits), with separate memory and central processing unit (CPU), stored program control, data entry via punched metal cards, and even an output printer. This ambitious device, though well specified, was never built.

The mechanical computing tradition was not entirely abandoned. Vannevar Bush built his analog mechanical computer, the Differential Analyzer, in 1930 at MIT. In 1954, M. Minsky and R. Silver used hydraulic logic elements to build a mechanical "hydroflip computer" which was operated at ~30 Hz and powered by a 3-inch high column of water describes the basis for a mechanical fluidic computer operating at ~5000 Hz.) In 1975, D. Hillis and B. Silverman built a special-purpose all-mechanical computer ~2 meters in size, entirely out of Tinkertoys, and powered by a hand crank, that was able to play tic-tac-toe. In 1990, University of Minnesota engineers fabricated a complete family of micromechanical digital logic devices, including electrostatically-actuated linear-sliding ~30 micron mechanical logic elements confined to a one-dimensional track, forming NAND and NOR gates suitable for low-speed radiation hard digital functions "in environments hostile to electronic devices." Sandia's pin-in-maze microlocks could also be used to make a mechanical computer, albeit at the above-micron scale. In 1996, J. Gimzewski and colleagues at IBM Zurich used a scanning tunneling microscope (STM) probe to repeatedly reposition spherical C_{60} fullerene molecules ~1 nm in diameter along a terraced copper substrate that constrained the buckyballs to move only in a straight line, operating the mechanical array like beads on an abacus. In 1997, Stoddart's group described a mechanical XOR gate based on their "molecular shuttles."

Perhaps the best-characterized (though not yet built) mechanical nanocomputer is Drexler's rod logic design. In this design, one sliding rod with a knob intersects a second knobbed sliding rod at right angles to the first. Depending upon the position of the first rod, the second may be free to move, or unable to move. This simple blocking interaction serves as the basis for logical operations. Drexler's benchmark mechanical nanocomputer design has 10^6 interlock gates, 10^5 logic rods, 10^4 registers, an energy-buffering flywheel and other components with total cubic volume ($\sim$400 nm)^3, mass $\sim 10^{-16}$ kg, and total power $\sim$60 nW, giving a power density of $\sim 10^{12}$ watts/m^3. Power dissipation per logic operation is $\sim$0.013 zJ per gate per cycle, giving (including register dissipation) $\sim 2 \times 10^4$ operations/sec-pW. Processing speed is $\sim 10^9$ operations/sec ($\sim$1 gigaflop) or $\sim 10^{28}$ operations/sec-m^3; assuming one bit per register, processing speed is $\sim 10^{13}$ bits/sec or $\sim 10^{32}$ bits/sec-m^3. In 1998, the typical desktop PC operated at $\sim 10^8$ operations/sec. Cooling is provided by a refrigerant fluid flowing through an integral fractal plumbing system at near the speed of sound.

The most primitive 4-bit Intel 4004 microprocessor, introduced in 1971 for early-generation pocket calculators, had 2300 transistors. A comparably simple early mechanical nanocomputer with 2000 interlock gates, 100 registers, and a 1 KHz clock speed could have a volume as small as 36,000 3 a ($\sim$33 nm)^3 cube, assuming 16 nm^3/gate and 40 nm^3/register, and could process $\sim 10^5$ bits/sec.

One major disadvantage of mechanical nanocomputers is that they will almost certainly be slower than nanoelectronic systems, because a device that depends on moving heavy nuclei ($\sim 10^{-27}$ kg) will necessarily be slower than a device that depends on moving less-massive electrons ($\sim 10^{-30}$ kg). Mechanical signals travel near the speed of sound ($\sim 10^4$ m/sec in diamond), whereas electronic signals may travel near the speed of light ($\sim 10^8$ m/sec). Thus while nanomechanical computers may be limited to switching speeds of $\sim$50 picoseconds, electronic devices may be $\sim 10^3$-10^4 times faster. On the other hand, mechanical computers are more EMP (electromagnetic pulse) resistant, conceptually easier to understand and to model, and mechanical designs scale readily from the macroscale to the nanoscale, unlike most nanoelectronic designs. In 1996, a $\sim$50 nm reversible mechanical latch was fabricated and physically cycled using an AFM; the similarity to Drexler's mechanical OR gate was explicitly noted in the paper.

What about mechanical data storage? In the MEMS world, Halg fabricated a nonvolatile microelectromechanical memory cell consisting of a longitudinally-stressed surface-micromachined $\sim$10 micron bridge that mechanically buckles into one of two states, making a bistable storage device. Moving to the nanoscale, data could be mechanically stored using compact three-dimensional arrays of diamondoid register rods (~10^7 bits/micron^3, ~10^{10} bits/sec access speed; and Drexler). Another theoretical mechanical storage medium is spooled hydrofluorocarbon memory tape (~10^{10} bits/micron^3, ~10^9 bits/sec access speed; and Drexler), or polymer or diamond surface-bound patterns of H and F atoms that could be read using an optimized $(CH_3)_3PO$ scanning probe tip with a maximum raw error rate of 6×10^{-8} per read. (Linear DNA molecules achieve ~10^9 bits/micron^3 data storage, by comparison.) Direct AFM reading and recording of 10-nm pits on a polycarbonate surface (~10^6 bits/micron, up to ~10^7 bits/sec access speed) and other means of directly reading and writing data at the atomic level have been studied, and the fullerene abacus described earlier is one of the first experimental efforts to mechanically store numerical information using individual molecules at room temperature.

It has been speculated that individual atoms or atomic vacancies could serve as information units. For example, Y. Mo used an STM to store information in the reversible rotational states of individual antimony dimers deposited on a silicon substrate. In theory, a block of diamond with data encoded as single-atom lattice vacancies could store up to ~176 bits/nm^3, or nearly ~2×10^{11} bits/micron^3, although cryogenic temperatures might be required to reduce diffusion effects and structural lability to acceptable levels consistent with long-term information storage. Phonon probes or other nondestructive means of lattice interrogation could be employed for readout.

Nanoelectronic Computers

In 1998, the metal-oxide-semiconductor field-effect transistor (MOSFET) was still the most common type of transistor in general use. These solid-state devices are characterized by an electrical source, an electrical drain, and a gate that controls the flow. Since its inception, the FET has scaled well to lower sizes. Individual working transistors with 40-nm gate lengths have been demonstrated in silicon, and 25-nm gate-length transistors have been fabricated in gallium arsenide. However, as feature sizes shrink below <~100 nm, a number of obstacles to stable operation begin to appear and the cost-effective downscaling of dense circuitry may not persist. A number of solid-state replacements for the bulk-effect semiconductor transistor have been suggested that could overcome these obstacles by taking advantage of nanometer-scale quantum mechanical effects. Such replacements include quantum dots (QDs), resonant tunneling devices (RTDs), electron turnstiles, and single-electron transistors (SETs). However, while hybrid RTD-FET circuits and SETs have been successfully switched at room temperature, many of these devices must be operated at cryogenic temperatures and other practical challenges remain; such top-down approaches to nanoelectronics are not considered further here.

In contrast, molecular nanoelectronics uses primarily covalently bonded molecular structures, electrically isolated from a bulk substrate, producing wires, switches and devices composed of individual molecules and nanometer-scale supramolecular structures. While it is relatively difficult and expensive to sculpt trillions of identical nanoscale structures in bulk materials (e.g., using 2-nm resolution electron-beam lithographic techniques), individual molecules are natural nanometer-scale structures that can be manufactured to be exactly the same in near-mole quantities. The great versatility of organic chemistry offers more options for designing and fabricating nanoelectronic devices than are available in silicon. Investigators are designing, modeling, fabricating, and testing individual molecules and nanometer-scale supramolecular structures that act as electrical switches and even exhibit some of the same properties as small solid-state transistors. For example, three and four terminal devices with ~0.3-nm feature sizes have been examined computationally.

Molecular electronics is a rapidly growing area in the literature. J.M. Tour expects experimental demonstration of molecular-sized transistor devices by 2000-2001, and commercial high-level computers using molecular electronics by 2008-2013. A current list of major research groups is maintained by the International Society for Molecular Electronics and Biocomputing. Some of the following discussion is appreciatively drawn from an extensive 1997 review by MITRE Corporation.

Molecular wires

Before one can seriously discuss molecular electronic devices embedded in single molecules, it must be asked whether a small single molecule can be made to conduct electrical current at all. A series of difficult and sensitive experiments and theoretical investigations during the 1990s answered this question affirmatively. Single molecule conduction was demonstrated by Tour using a conducting molecule characterized by repeating structures—in this case, a series of benzene-like rings connected by acetylene linkages—each part of which is linked to the next by bonds including many p-electrons above and below the plane of the structure. These electrons in the p orbitals conjugate with each other, or interact, to form a single large orbital throughout the length of the wire to permit mobile electrons to flow. Thiol (-SH) functional groups at either end adsorb well to gold surfaces and act as "alligator clips" for attaching molecular electronic units to metal substrates. Tour's polyphenylene wires can carry ~nanoampere currents. Such molecular wires also have the desirable property that they can be made quite long, if necessary, because they can be lengthened systematically using chemosynthetic methods. Another experiment using a single benzene ring attached between two gold electrodes through thiol groups across a 0.846-nm gap found a singlering threshold resistance of ~20 megohm and a capacitance on the order of ~0.1 e/kT ~ 10^{-19} farad.

Fullerene carbon nanotubes or "buckytubes" are hollow cylindrical tubes, essentially rolled-up sheets of graphite, measuring > ~1 nm in diameter that (in many chiral and compositional forms can conduct electricity. For instance, both individual 10-20 nm diameter carbon nanotubes and atomic wires pulled from an unraveling carbon nanotube can conduct currents of ~1-10 microamps (vs. nanoamps via tunneling). Maximum nanotube current density is ~10^{10}-10^{11} amp/m over a 0.2-6.0 volt range, superior to the ~10^9 amp/m typically achieved in superconductors. Nanotube resistivity measures ~8-20 ohm-m for straight tubes, ~38-49 ohm-m for slightly curved (5-30°) tubes, and >100 ohm-m for highly curved (65-80°) tubes. Interestingly, electronic transport in carbon nanotubes is ballistic (e.g., like electron waveguides, permitting only a few propagating modes), with all waste heat being dissipated in the leads to the nanotube element and none in the nanotube itself. Buckytubes are stiff enough to serve as supports for molecular circuit elements, and if filled with conducting metal can create one of the structurally strongest nanowires that is chemically possible.

Other 0.6-3 nm molecular wires, 3-20 nm semiconductor wires, and nanoscale metallic wires have been studied as well.

Electromechanical devices

In 1998, the best-studied category of molecular electronics device is probably the electromechanical molecular switch. This class of switch employs electrically or mechanically applied forces to turn a current on and off, either by changing the conformation of the molecule or by moving a switching molecule or group of atoms in the manner of a gate. These switches are promising because they could be laid down in a dense network on a solid substrate, or in some cases in three-dimensional arrays; unfortunately, no integrated CPU-scale designs have yet been attempted, although plans for a half-adder and a molecular shift register have been published. Some examples:

Electromechanical amplifier

Joachim and Girnzewski measured conductance through a single buckyball held between an STM tip and a conducting substrate. By pressing down harder on the STM tip, the buckyball deforms, tuning the conduction off-resonance and reducing current ~50%. Although the STM tip could be replaced with a small in-situ piezoelectric gate, the ideal actuator would be a molecular-scale electromechanical actuator akin to a logic rod; the major fundamental limit to the speed of this switch is the natural mechanical vibrational resonance frequencies of a buckyball, ~10^{13} Hz.

Linear atomic relay

Researchers at Hitachi Corporation simulated a two-state electronic switch wherein a mobile atom that is not firmly attached to a substrate moves back and forth between two terminals. If the switching atom is in place, the device conducts electricity; if the switching atom is displaced from the two wires, the resulting gap greatly reduces the current that can flow through the atom wire. A small negative charge placed on the third atom wire "gate" near the switching atom moves the switching atom out of its place in the wire; the switching atom is pulled back into place by a second "reset" gate after each use of the switch. Actual experiments approximating this design created a bistable atom switch using a xenon switching atom moving back and forth between an STM tip and a substrate, demonstrating that the movement of a single atom can be the basis of a nanometer-scale switch. In a more mature device, relays could be ~10 nm in size with a speed limited only by the intrinsic vibrational frequency of atoms, typically ~100 THz, but it seems likely that atom relays could only operate at very low temperatures because not much energy would be required to evaporate a switching atom off the substrate and out of the plane of the atom wires, destroying the switch.

Rotational molecular relay

A more reliable two-state device based on atom movement uses the rotation of a molecular group to affect an electric current. The switching atom is part of a rotating group, or "rotamer," which itself is part of a larger molecule, perhaps affixed to the same surface as the atom wires. Alternatively, hydrogen bonding might provide enough rotational resistance to stop the rotamer in the conducting position, but not so much that reversing the gate voltage would be insufficient to turn the rotamer. Controlled rotational state switching of a Pt-adsorbed oxygen molecule by STM using 0.15-volt 20-microsec pulses was demonstrated in 1998.

While a rotating switch with a methyl-like rotamer group has three distinct switch positions, a hinging switch that rotates back and forth between two distinct states might be preferable. Cyclohexane, a simple example of this kind of molecule, can bend into two different forms, commonly known at the "boat" and "chair" conformations. The cyclohexane-type molecule attaches to a molecular framework while the remaining ring carbons are replaced by groups tailored to use steric repulsions or van der Waals attractions to reduce undesired switching caused by thermal energy. Switch speed is limited by molecular rotational and torsional frequencies, typically ~1-1000 GHz, which is slower than the atom relay but more reliable. Also in contrast to the atom relay, molecular relays could be packed in three dimensions, possibly up to ~1 nm^3 densities.

Molecular shuttle switch

A rotaxane "shuttle switch" has been synthesized by F. Stoddart's group that consists of one ring-shaped molecule that encircles and slides along a shaft-like chain molecule. Two large terminal groups at either end of the shaft prevent the shuttle ring from slipping off the shaft. The shaft contains two other functional groups, a biphenol group and a benzidine group, that serve as natural stations between which the shuttle moves. The shuttle molecule contains four positively charged functional groups that cause it to be attracted to shaft sites with extra negative change, so the shuttle spends 84% of its time at the benzidine station, which is a better electron donor than the biphenol station. Removing an electron from the benzidine station forces the shuttle to switch to the other station; switching may be controlled chemically or electrochemically. A charged gate could be added at one or both ends to force the shuttle to move between stations, with switch state probed by arranging for the ring to complete an electric circuit in one of its two positions. Switching speed is slow, limited by the speed of electron transfer to and from the benzidine station and the sluggish motion of the ring which is ~10^6 times heavier than an electron, but three-dimensional packing densities of ~0.01 nm^3 appear plausible. Other rotaxane shuttle systems have been studied. In 1999, the first chemically-assembled rotaxane-based logic gate was demonstrated, although the gates could be opened only irreversibly.

Field-controlled molecular devices

A second category of molecular electronics device that appears promising for the development of molecular nanoelectronic digital computers is the electric-field controlled molecular switching devices, including molecular quantum-effect devices. These are most closely descended from the solid-state microelectronics and nanoelectronic devices described earlier, and promise to be the fastest and most densely integrated among all of the current alternatives. A group at Purdue University has used self-assembly to fabricate and demonstrate functioning arrays of molecular electronic quantum confinement structures connected by molecular wires. As another example, J.M. Tour proposes embedding a quantum well in a molecular wire by inserting pairs of barrier groups that break the sequence of conjugated porbitals forming a two-terminal molecular RTD; other structures for three-terminal molecules have been suggested. (In February 1999, Tour reported the demonstration of RTD effects in the manner proposed above; at press time, the paper was still in review. Nondissipative (adiabatic) Thouless electron pumping has also been described at 0.33 K.

Merkle and Drexler consider a hypothetical "helical logic" device in which a single electron can switch another single electron, with 1 or 0 represented as the presence or absence of individual electrons. The electrons are constrained to move along helical paths, driven by a rotating immersive electric field (normal to the helical axis) that confines each charge carrier to a fraction of a turn of a single helical loop, moving it like water in an Archimedean screw. A logic operation involves two helices, one of which splits into two "descendant" helices. At the point of divergence, differences in the electrostatic potential resulting from the presence or absence of a carrier in the adjacent helix control the direction taken by a carrier in the splitting helix; the sequence is reversible, allowing two initially distinct helical paths to merge into a single outgoing helical path without forcing a dissipative transition. Energy dissipation at ~10 GHz is ~10^{-6} zJ/gate-cycle at an operating temperature of 1 K.

Other molecular electronic devices

Other classes of molecular electronic devices have been studied, including picosecond photoactive or photochromic molecular switching devices using light, and electrochemical molecular devices using electrochemical reactions, to change the shape, orientation, or electron configuration of a molecule in order to switch a current. However, photoactive devices in a dense network would require an NSOM-like emitter to switch individually, since optical photon pathways are not easily confined at length scales very much below optical wavelengths of ~500-1000 nm; electrochemical molecular devices might require immersion in solvent to operate.

Fullerenes exhibit a wide range of quantized conductivities that may be useful in nanoelectronic computers. For example, chemisorption alters properties: 3 alkali dopant atoms per C_{60} buckyball gives high conductivity, while none or 6 dopant atoms gives an insulator. Buckytube conductivity also varies with mechanical stress and torsion, tube diameter, and other geometrical parameters. Depending upon the direction of the nanotube chiral vector, carbon nanotubes may be either metallic or semiconducting. Suitably-sized, doped, flexed, stressed, or chiralized fullerenes could be nested to make insulated wires or joined to make nanoelectronic computer components including resistors with quantized "staircase" resistance of nh/2e, n = 1, 2,... (e is electronic charge, h is Planck's constant), two-terminal devices such as diodes, three- or four-terminal devices such as heterojunction, TUBEFET, or thin-film transistors, and gates that introduce electric fields. A single-molecule transistor using semiconducting buckytubes has been demonstrated experimentally at room temperature.

Carbon nanotubes have been cut to length, size-sorted by field flow fractionation, their open ends derivatized with thiol groups and then tethered to 10-nm gold particle "junction boxes" which AFM images show can connect together at least two separate tubes. If the tethering process can be spatially controlled using DNA complementarity or by other means, then fullerene nanoelectronic circuit elements could be assembled into three-dimensional CPU-like structures. P. Collins expects an all-carbon 8-bit adder with nanometer dimensions by the year 2002, and J.C. Ellenbogen of MITRE Corp. has designed a molecular half-adder that measures ~10 nm × 10 nm in size. This research area was extremely active in 1998.

Fullerenes may also allow compact memory devices. For example, it has long been speculated that a ~1 nm buckyball encapsulated in a ~1 nm diameter nanotube segment may glide freely back and forth, trapped weakly in each end cap by van der Waals forces. An external voltage could nudge a charged C_{60} molecule to one end of the other, creating a two-state "buckyshuttle RAM" storage device. Such a device could store one bit of information in a ~10 nm^3 volume, a storage density of ~10^8 bits/$micron^3$; the C_{60} thermal velocity of ~100 m/sec at 310 K inside a ~10 nm memory element gives a typical read/write time of ~0.1 nanosec. In early 1999, Kwon and colleagues reported that finely dispersed 4-6 nm diamond powder thermally annealed by heating in a graphite crucible under argon at 1800°C for 1 hour produced oblong multiwall carbon nanostructures which in some cases could move around inside each other, like C_{60} trapped within C_{480}. Kwon noted that there are potential energy minima at either end, so a charged buckyball (e.g., an enclosed ion) could be shuttled with an applied electric field, which could be used for memory storage. Computational simulation of writing a bit showed that the interior buckyball neatly shuttled to the correct end, then gently bounced to rest in ~0.01 nanosec (i.e., ~100 GHz operation) dissipating only ~10 kT in the C_{480}. Kwon proposed a high density memory board using aligned buckyshuttles in a hexagonal lattice with addressing wires above and below, not unlike a ferrite core memory; the memory would likely be nonvolatile at room temperature, with mass production based on self-assembly.

Electrostatic field computers

In 1998, bulk semiconductor technology typically required ~16,000 electrons to transfer one bit of information. SET and related techniques described above promise to reduce this to ~1 electron per bit. But a "traditional" molecular electronics nanocomputer with a mean gate density of ~1 nm^{-3} operating at ~10 GHz and using single electrons to transport, indicate, fetch, or represent a binary digit would move ~1 coul/sec of electrons (~1 amp at ~1 volt, or ~1 watt) through a 1 $micron^3$ CPU, producing an untenable ~10^{18} $watt/m^3$ power density, compared to ~10^{12} $watts/m^3$ for the nanomechanical CPU described in.

In order to reclaim this factor of 10^6, Tour and Seminario conclude that a molecular CPU must operate by electrostatic interactions rather than by electron currents, using intermolecular "talk" to efficiently transfer a bit using only ~10^{-6} electron. This charge-density approach is already well-known in biology, where enzyme-ligand recognition is mediated by hydrogen bonding or van der Waals interactions, working through the electric fields of the molecules even though no formal charge transfer occurs. Similarly, in an electrostatic nanocomputer the input signal is initiated by bringing up a charge to one end of the molecular device, thus reshaping the local charge density. That molecular device, in turn, perturbs the electrostatic field of an adjacent molecular device, rearranging its electron density in $< 10^{-15}$ sec, and so the electrostatic field perturbation is propagated without charge transfer: "These movements of electrons (wave-particle entities) in the molecular orbitals (or circuits) are performed without any dissipation of energy (stationary states)." Perturbation pulses may have an energy magnitude similar to those of van der Waals interactions. The hypothetical electrostatic molecular logic device shown in serves as an OR gate if the output is active (as a potential) or as a NOR gate if the output is passive (as an impedance) if positive logic is used, or an AND or NAND gate if negative logic is used.

Gates at input/output interfaces must still be able to receive or to drive signals from standard electronic circuits, but the vast majority of molecular gates would lie within the molecular CPU and could process information by controllably affecting the electrostatic potential of neighbouring molecules. Such devices would likely shift the hardware/software equilibrium towards hardware, with massively wired-logic supplanting programmed-logic computing, and with large molecular arrays allowing computation to take place inside the CPU with minimal need for main or auxiliary memories. Tour expects the first laboratory demonstration of electrostatic field information transfer by 2001-2003. J.C. Ellenbogen notes that a charge density nanocomputer achieves low dissipation by operating very close to equilibrium at all times, but notes that clock speeds of at least 1-10 KHz should still be obtainable with this approach; Tour claims that 1-10 THz switching is possible. Current-free polarization gates have also been proposed and fabricated.

Biocomputers

Biocomputers use natural organic materials, thus offer the possibility of self-assembly or construction by natural or engineered biological organisms. At least three classes of biocomputers are distinguishable—biochemical, biomechanical, and bioelectronic.

Biochemical computers

The Systron-Donner Analog Computer familiar to engineering students in the 1960s and 1970s used potentiometers and high-gain DC amplifiers, called *operational amplifiers*, to allow the creation of electrical circuits comprised of adders, multipliers, and integrators arranged in feedback loops capable of solving complex time-dependent differential equations modeling actual physical systems. In 1998, Motorola and others were producing programmable analog chips containing arrays of operational amplifiers and other analog components. Related control systems are commonplace in human physiology.

Biochemical feedback loops exemplified by enzyme activity in the living cell exhibit similar analog computational functionality. To start with, the net flow of carbon through an enzyme-catalyzed reaction may be influenced by

1. the absolute quantity of enzyme present,
2. the pool of nonenzyme reactants and products available, and
3. the catalytic efficiency of the enzyme.

All three influences may themselves be subject to recursive enzymatic control. For instance, the absolute quantity of enzyme present is determined by competing rates of synthesis and degradation; both of these processes are controlled by other enzymes. Product synthesis may be triggered by an inducer, or may be repressed by an end product. Local concentrations of coenzymes or metal ions can regulate catalytic efficiency. Inhibited activity of an enzyme in a biosynthetic pathway by an end product of that pathway, such as the much-studied inhibition of aspartate transcarbamoylase by cytidine triphosphate (CTP), is called *feedback inhibition*.

Typically the feedback inhibitor, acting as a negative allosteric effector, binds to the sensitive enzyme at an allosteric site that is spatially distinct from the catalytic site. Further fine control is provided by additional multiple feedback loops, exhibiting such features as cumulative feedback inhibition (e.g., the inhibitory effect of two or more end products on a single regulatory enzyme is strictly additive), multivalent feedback inhibition (e.g., complete inhibition occurs only when two or more end products both are present in excess), and cooperative feedback inhibition (e.g., single end product present in excess inhibits the enzyme, but inhibitory effect of two or more end products present in excess is "multiplicative"). Similarly, parallel chemical computers consisting of open, bistable coupled reaction systems capable of pattern storage and pattern recognition have been proposed and simulated to model a neural network, and experiments have been performed with a system of 16 coupled bistable chemical oscillators using the chlorite-iodide reaction.

Networks of interacting enzymatic reactivities can also be used to create biochemical digital computers. An enzymatic Turing machine was first proposed by C.H. Bennett, and Liberman described the living cell as an analog-digital molecular stochastic parallel computer. More recently, T. Knight of the MIT Artificial Intelligence Laboratory suggested that normal cellular activities can be coopted to implement digital logic gates, a process he calls "*cellular gate technology*." In this scheme, protein concentrations are used as binary signals. The signal is "1" if the marker protein is being synthesized, thus reaching some detectable threshold equilibrium value where the rates of synthesis and degradation are equal. The signal is "0" if the synthesis of the marker protein is being inhibited by some other protein, and thus the concentration is trending toward zero.

This principle may be used to make logical circuit elements. For example, an inverter system consists of two different proteins, one of which is inhibiting the synthesis of the other. A NOR gate requires a system in which either of two proteins can inhibit the synthesis of a third protein; multivalent feedback inhibition in which two proteins are both required to inhibit production of a third makes a NAND gate. This is sufficient to implement any Boolean function, given enough gates and signals. Two NORs crossed in a loop makes a bistable multivibrator, or flip-flop, a fundamental component in data registers, shift registers, rectangular pulse generators, or pulse delay elements. The logic circuitry (the arrangement of "gates" and "wires") comprises a set of designed DNA sequences that determine which proteins are being synthesized and which combinations of proteins can inhibit the synthesis of others.

In the most optimistic case, we assume operation near concentrations of $c_{prot} \sim 10^{-4}$ gm/cm^3 (similar to most complement factors and some clotting factors in human blood; marker proteins and gating enzymes of mass $MW_{prot} \sim 10$ kilodaltons, and a minimum of $N_{prot} \sim 100$ protein molecules per gate, giving a typical logic element size of $V_{gate} \sim N_{prot}\ MW_{prot} / N_A\ c_{prot} \sim 0.02$ micron^3/gate where N_A is Avogadro's number. Flip-flop registers may store ~50 bits/micron^3. In theory, the DNA specifications for this computer could be grafted onto the genome of an animal cell (~20 micron^3 = up to 1000 gate volumes) or a bacterial cell (~2 micron^3 = up to 100 gate volumes), taking care to avoid any unwanted interactions with natural biochemical pathways. Minimum diffusion-limited gating time across a biochemical gate of characteristic dimension $DX \sim V_{gate}$ 1/3 is given by as $t = DX^2/2\,D \sim 0.3$ millisec assuming protein marker diffusivity $D \sim 10^{-10}$ m/sec, consistent with ~KHz maximum operating frequencies when DNA is localized near the gates. For nucleoplasmic DNA with cytoplasmic computation, diffusion-limited maximum switching frequency falls to ~0.1-1 Hz. For comparison with natural systems, genes controlling the fibroblast cell proliferation program have been observed to switch states in ~900 sec, implying computation at ~0.001 Hz.

Other classes of biochemically-modulated biomolecular switches have been investigated. For example, Matthews incorporated an artificial molecular on-off switch into an enzyme by adding a pair of thiol groups into the active site of native lysozyme, replacing two amino acids on opposite side of the active site with cysteine, an amino acid with a thiol-containing side chain. Thiols may form covalent bridges (-SS-) or broken bridges (-SH HS-) under suitable chemical conditions. Cycling the solution composition causes the cysteines to form a bridge across the active site (inactivating the enzyme) or to break the bridge (activating the enzyme); the process is completely reversible. Another experiment involved genetic modification to the bacterial phage lambda, a virus with two behavioural states called lytic (replication-active) and lysogenic (replication-inactive).

The NIH researchers installed a synthetic molecular switch in lambda that allows the virus to be biochemically toggled between its lysogenic and lytic phases by the introduction of active HIV-1 protease. The genetic circuit that controls the phage lambda lysis-lysogeny decision has been flow-charted in detail using Boolean logic operators. The immunological synapse—the specialized junction between a T cell and an antigen-presenting cell —acts as a kind of biochemical logic gate with a ~ 0.06 Hz switching rate.

In 1998, the DNA molecule also was being actively investigated for biochemical computers. DNA-based Turing machines were first proposed by C.H. Bennett and R. Landauer. In 1994, L. Adelman demonstrated the first DNA computer using fragments of DNA to compute the solution to a simple graph theory problem. Adieman used short DNA sequences to represent vertices of a network, or graph. Combinations of these sequences were then synthesized randomly by massively parallel chemical reactions in aqueous solution using a combination of the copying and combination reactions applied to the artificial DNA strands, comprising all possible random paths through the graph, after which a sequence representing the designed result was biochemically extracted from the stew.

Initially it was thought that DNA computing would be limited to the solution of combinatorial problems, but subsequent research has shown that the approach can be applied to a much wider class of digital computations. The well-known problem of combinatorial explosion (e.g., even a small protein of 300 amino acids has a sequence space of 10^{390}) leading to huge-volume DNA libraries during a complex computation may be avoided using a recursive selection approach crudely analogous to genetic algorithms. Some work has been done using plasmids and restriction endonucleases to implement a DNA-based Turing machine, using only commercially available restriction enzymes and ligases for every operation with states represented as sequences of bases around a plasmid, and Warren Smith has another Turing machine design using the chemistry of guide RNAs in Trypanosome kinetoplasts. Practical challenges include finding fast, efficient, and low-noise input and output techniques.

DNA computation is in theory energetically efficient, requiring only ~12 kT per ligation operation at room temperature, with massively parallel operations although each ligation reaction requires ~1 sec to complete. DNA data representation allows up to ~1 bit/nm^3 storage density. However, a practical mass storage device would probably require 10-100 nucleotide-long words to avoid ambiguities during the recall process, thus reducing maximum density to 10^7-10^8 bits/micron^3, though of course some of this reduction could be offset by reducing read rate requirements (e.g., by using multiple read heads with majority logic).

Biomechanical computers

Components of mechanical computers could be constructed entirely of biologically-derived materials which in theory might be fabricated using synthesis pathways accessible to engineered microbiota. For example, Drexler's mechanical nanocomputer assumes diamondoid logic rods with a Young's modulus of $E = 5 \times 10^{11}$ N/m, but the Young's modulus for human bone hydroxyapatite and fluoroapatite crystals is almost as high, with $E \sim 1\text{-}2 \times 10^{11}$ N/m. Ignoring the significant problems inherent in noneutactic rod logic systems, apatite-based mechanical computer structures might be laid down using engineered osteocytes, perhaps assisted by chemotactic or contact guidance techniques. Failure strength of apatites is ~100 times poorer than for diamond, so apatite logic rod force and acceleration should be reduced by ~100 and rod vibrational energy by ~10^4, increasing minimum switching time by a factor of ~10 (to ~1 nanosec) and decreasing maximum rod speed to ~1 m/sec.

N. Seeman has constructed molecular building blocks from unusual DNA motifs, using stable-branched DNA molecules with the connectivity of a cube or a truncated octahedron. E. Winfree has proposed using arrays of DNA crossover molecules in DNA-based computing, requiring the ability to build periodic backbones with bases differing from one unit cell to another. In addition to branching topology,

DNA also allows control of linking topology. DNA-based topological control has led to the construction of Borromean rings which could be used in DNA-based computing applications. In Borromean rings, the linkage between any pair of rings disappears in the absence of the third. Rings can be designed with an arbitrary number of circles; the integrity of a link could represent the truth of each of a group of logical statements.

DNA structural transitions could be used to drive nanomechanical devices via branch migration. Application of torque to a cruciform leads to the extrusion or intrusion of a cruciform. A synthetic branched junction with two opposite arms linked can relocate its branch point in response to positive ethidium-induced supercoiling, representing the first experimental step in developing DNA structural transitions that can achieve a nanomechanical result. The possible use of the B-Z transition (e.g., from right-handed B-DNA to left-handed Z-DNA) in nanomechanical devices is being explored.

In 1998, Winfree and colleagues in Seeman's laboratory described a simple, predictable, highly precise technique for arranging DNA molecules into two-dimensional crystals. At the same time, a paper by Heath and others described a defect-tolerant computer architecture called the Teramac, a massively parallel experimental computer which apparently could operate normally with at least half of its most critical components failed. The authors claimed this architecture showed it would be "feasible to chemically synthesize individual [molecular] electronic components with less than a 100% yield, assemble them into systems with appreciable uncertainty in their connectivity, and still create a powerful and reliable data communications network."

One can also imagine a network of interlocking enzyme-like conformational switches, possibly embedded in a semirigid two- or three-dimensional scaffolding of DNA or protein molecules. Enzymes may display multivalent feedback inhibition (making a NAND gate; or may have multiple metal-ion binding sites influencing enzymatic action; an enzyme requiring two conformational changes to inhibit its action could serve as a biomechanical NAND gate if tethered to other similar enzymes in an appropriately structured protein array. Single-device error rates will be high but might be reduced to acceptable levels by employing short multiply-redundant pathways for each digital computational operation, or a sufficiently parallel architecture. Site-directed mutagenesis optimization of antibodies has shown that natural proteins can be made significantly stiffer and more stable by the mutation of just a few amino acids. Systems of reciprocating actomyosin molecular motors (e.g., as found in muscle fibers; might serve as interdigitating components of a push-pull mechanical logic architecture resembling rod logic or a three-dimensional loom. In early 1999, Viola Vogel and colleagues discovered, via computer simulation, a tension-activated fully-reversible biomechanical switch comprised of a single strand of fibronectin protein—like untying a shoelace, a slight mechanical tug unravels a folded segment, switching off the protein's biochemical activity.

Mechanical coupling of intracellular Ca^{++} ion release channels allows coordinated voltage gating across the surface membrane of the sarcoplasmic reticulum inside muscle cells. R. Bradbury suggests that site-directed mutagenesis could produce a family of proteins that could be activated by the release of various mono- (e.g., K^+), di- (e.g., Ca^{++}), and trivalent (e.g., Al^{+++}) ions. This would allow multivalued logic (e.g., K = 1, Ca = 2, Al = 4), or increased parallelism or bandwidth (e.g., Ca = !(K & K), Al = !(Ca & Ca), etc.); once a protein exists that can be activated by a single ion, then varying its architecture to allow substitution of an ion of different size or charge is a relatively straightforward engineering exercise. A biomechanical Turing machine design has also been reported by Shapiro.

Bioelectronic computers

Many organic materials may be useful in electronic computing. Electrical conduction in synthetic organic molecules is well known and already exploited in liquid-crystal displays of laptop computers. Doped polyacetylene is a conducting plastic with a carbon backbone; undoped, it is an insulator. There are many other known families of conducting polymers such as polyaniline, which includes nitrogen atoms in the backbone, and polythiophene, which includes sulfur. Three-dimensional interconnect arrays of organic conducting polymers have been fabricated; conductivity is higher along a chain than between chains, so sheets of oriented chains can produce relative conductivity anisotropies as high as 100-1000. An all-plastic transistor has been built.

Organic transistors have been used as the active semiconducting element in thin-film transistors fabricated with organic-material thicknesses ranging from 5-150 nm, and organic microscale transistors and other organic devices may exploit bulk-effect electron transport just like silicon-based semiconductor devices. Conjugated polymer poly(1,6-heptadiester) was employed to make an optical correlator with a peak processing rate of 3×10^{16} operations/sec. Spin-transition polymers could serve as thermal sensors or memory devices with the theoretical ability to store one bit in a ~4 nm cube (~2×10^{7} bits/micron^3 storage density) with ~GHz addressing speeds at room temperature. Other organic polymers possibly useful in logic circuits have been investigated.

Natural biological materials may also be useful in electronic computing, particularly because of the ability of biological materials to self-assemble. Biomolecular electronics is a subfield of molecular electronics that investigates the use of native as well as modified biological molecules (chromophores, proteins, etc.) in place of the organic molecules synthesized in the laboratory. In theory, a three-dimensional DNA or protein scaffolding could be self-assembled, with bacteriorhodopsin, ferritin and magnetite, or related bioelectromagnetic molecules inserted into precise locations within the structure to produce a crystal lattice bioelectronic or biooptoelectronic nanocomputer. DNA already has been used as an atomic scaffold upon which to build silver wires ~100 nm in diameter and has been decorated with fullerenes. Self-assembly can generate membrane-based and biotubule-based devices, and site-directed mutagenesis is a valuable tool for high-resolution protein device engineering.

Nucleotide sequence-specific electron transfer between metallic electron donor and acceptor complexes covalently or noncovalently intercalated into strands of B-DNA has been demonstrated. Perhaps the best-known bioelectronic (and biochemical) computer is the neuron cell and its congeries—the ganglia, nerve trunks, and brains. In 1997, David Stenger (NRL) and James Hickman (SAIC) were attempting to culture and link together neuronal cells to build a bioelectronic cell-based sensor "computer" for performing complex pattern-recognition tasks. In early 1999, W.L. Ditto at Georgia Tech announced the creation of a biological computer comprised entirely of leech neurons, that was capable of performing simple sums. We can certainly imagine large numbers of cultured neurons arranged in artificial three-dimensional spatial patterns to produce a synthetic bioneural computer, but such a computer would operate at only ~KHz frequencies and would be very energy inefficient since each neuron consumes ~10^{10} kT per discharge—thus offering few advantages.

Ultimate Limits to Computation

A nanosystem designer quickly discovers that total nanorobot power demand is often dominated by computational heat dissipation, rather than by energy dissipation from chemical or mechanical processes. Thus important design objectives are to:

1. minimize onboard computation,
2. maximize algorithmic efficiency (e.g., super-Turing computers), and
3. minimize power dissipation per computation operation, an objective which may be met using reversible computing.

It is also interesting to consider the maximum possible information density in physical matter as a calibration for current and projected achievements.

Reversible computers

Computers may be thought of as engines for transforming free energy into waste heat and mathematical work. Early pioneers of computing theory believed that each step in a computer's binary computation required a minimum energy expenditure of ~kT ln(2) ~3 zJ/bit at T = 310 K. In 1961, R. Landauer argued that it was the erasure of information, not computation per se, that generates waste heat. It is now known that computers can in principle do an arbitrarily large amount of reliable computation per kT of energy dissipated. Following Landauer's insight, Fredkin and Toffoli suggested an idealized "ballistic computer" that could, in theory, compute at finite speed with zero energy dissipation and zero error. A more pragmatic family of models are the "Brownian computers" in which thermal noise pushes system elements in a random walk throughout the entire accessible portion of the computer's configuration space; in these models, energy dissipation trends to zero only in the limit of zero speed. Both ballistic and Brownian computers require that all computations are logically reversible, with no irreversible bit erasures and no machine state having more than one logical predecessor—that is, the output uniquely specifies the input.

One simple implementation of reversible computing is the retractile cascade. In a retractile cascade, all inputs and all intermediate states leading to a result are retained during the course of a computation. After the computation is complete, the final result is copied to an output register, requiring the irreversible erasure of only enough bits in the output register to hold a copy of the final result, which need cost no more than kT ln(2) per erased bit. The computation is then reversed, step by step, culminating with the original inputs; the slower the reversible steps are performed, the less energy they may dissipate (but the longer the computation takes). Schematics for a retractile

AND gate, adder, shifter, and programmable logic array have been published, and in 1998 a reversible processor based on a modern RISC architecture was being designed. A conventional computer architecture, implemented without regard to reversibility, may perform 0.1-1 bit erasures/gate-cycle; by comparison, a retractile computer might average $<10^{-4}$ bit erasures/gate-cycle. During the reversible portion of the computation, Drexler's exemplar rod logic design employs a retractile cascade that reduces room-temperature energy dissipation from the "classical" minimum of ~0.7 kT down to ~0.003 kT per gate-cycle. In the limit of slow motion, all identified energy dissipation mechanisms in combinatorial rod logic systems approach zero. The Tour-Seminario electrostatic field switch might attain ~10^{-5} kT per gate-cycle at room temperature; low-temperature helical logic could achieve ~10^{-7} kT per gate-cycle. Feynman notes that the minimum free energy required for a reversible computation may be made independent of the complexity or number of steps in the calculation, and may be as small as ~kT per bit of the output answer.

J.S. Hall suggests two principal design rules for efficient nanocomputers:

1. erase as few bits as possible, and
2. eliminate entropy loss in operations that do not erase bits.

Quantum computers

Digital computers manipulate discrete units of information, with data strings stored as binary digits, or bits. Any two-state physical system (e.g., high or low voltages in semiconductor circuits) can represent these bits. Similarly, a quantum computer would use quantum states of atomic or molecular systems to store data. For example, the states of a hydrogen atom could be assigned such that a wave function corresponding to a state of hydrogen would represent a bit of data. Shining laser light on the atoms would trigger processing the data by inducing transitions between electronic states.

Quantum systems, however, exhibit a peculiar phenomenon known as superposition, in which several discrete states can be processed by a single physical system at once, crudely analogous to the musical effect of a single note being made up of many different harmonics. Superposition gives quantum systems the potential to be enormously more powerful than conventional computers. Since more than one state can be supported, quantum bits, or qubits, may store a mixture of many bits of data at the same time. A superposition of N qubits can store 2^N binary digits. An operation on these mixtures is massively parallel, effectively performing many calculations simultaneously. For example, photons interacting with an atom in a superposition of states drive all the states in the superposition, producing another superposition corresponding to all solutions of the original states. The advantage over conventional computers is not merely linear, but exponential: A conventional processor operates sequentially on 64-bit numbers, but a quantum computer with 64 qubits would simultaneously operate on the full set of 2^{64} ($\sim 10^{19}$) binary values. An N-qubit quantum computer could model an N-body system in real time; in contrast, the number of operations required by a conventional computer to perform such a simulation increases exponentially with the number of bodies.

Since the early 1980s, theoretical interest in quantum computing has developed rapidly. Many groups have begun designing quantum computational architectures and constructing physical quantum gates and devices. In 1995, one group at NIST made a controlled NOT gate of two qubits from a cooled beryllium ion in a radio-frequency trap, while a French group used single photons trapped in quantum cavities to control cesium atom states. In 1999, experiments began on an analog quantum computer using electrons floating on liquid helium. Others work with NMR techniques, storing data not in electron states but in nuclear spins, which are less susceptible to perturbations. Resonance effects between proton and electron spins in hydrogen atoms can make AND and NOT gates, hence NAND gates, from which all Boolean computers can be made. It has been suggested that by encoding the amplitudes of a couple of thousand electron states in a superposition, large amounts of data could be written onto a single atom. The discovery of certain error-correcting procedures, since verified experimentally, was another important breakthrough. Unlike conventional devices, it is not possible to simply check that qubits are in the right states, because the act of measuring destroys the coherence. But Shor proved that information can be reliably stored and read from a several qubit system even if one of the bits is corrupted, and it has been shown that error correction can be applied during the computation process itself. This opens the door to many-qubit systems; by 1998, two groups had already performed 4-bit sums, and one group had demonstrated a 2-qubit device. Quantum game theory has also been investigated.

Dedicated Computational Organs

Nanostructured artificial organs specializing in control, information processing, or data storage could be permanently embedded within the human body. Such organs could provide a user-accessible computational facility, or help coordinate and control (or even subsume) the activities of in vivo medical nanorobots or other dedicated nano-organs, including energy, communications, navigation, or chronometer systems. Computational organs could also serve as embedded data repositories (library nodules), including personal medical and experiential records, customized calendars and personal data banks, or as communications encryption and security interfaces.

In 1998, microchips glued to honeybees or implanted under the skin of livestock and household pets contained owner/address information that could be retrieved by passing a reading wand over the animal (e.g.. Pet Trac, manufactured by A.V.I.D. Identification Systems; Norco, California). Read-only microchips with identification numbers for individual cattle were encased in acid-resistant porcelain and inserted into an animal's stomach ten days after birth. These chips could be read by a handheld computer or a stationary reader; Shearwell Data, which sold the chips for $6 each, planned to adapt them to communicate with global positioning satellites so that farmers could remotely track herd movements. Peter Cochranenotes that a colleague has had a computer chip embedded under his skin, bearing such personal information as medical records and bank account and passport numbers. There were also unconfirmed rumors of a subdermally-implanted bio-powered (4 milliamp) 16 mm^2 transmitter chip called Sky-Eye, sold by Gen-Etics for $7,500 as kidnap deterrence for "the rich and famous," that supposedly emitted a personal tracking signal detectable by satellite.

Cytocide and Virucide

The objective of nanomedical treatment is often cell repair or replacement. However, it will frequently be necessary for medical nanorobots to destroy foreign viruses and pathogenic bacteria, protozoa, or metazoan parasites. It may also become necessary to eliminate native human tissue cells that have grown too numerous in a local region, cells that have adopted or developed major pathological, teratological or other undesirable patterns, or which have been too severely traumatized to permit efficient repair.

There are at least two classes of nanomedical cytocidal instrumentalities that may be employed to rid the human body of unwanted cells. The first may be termed the biochemical approach, including methods which may take up to hours or days to proceed to completion with acceptable reliability. Biochemical methods can require relatively modest technological sophistication to implement—an important consideration in the early years of nanomedical device development. The second method may be termed the mechanical approach, which may require only seconds or minutes to proceed to completion with very high reliability, but which involves more sophisticated nanomechanisms that might only be available in a mature nanomedical technology environment.

Biochemical cytocide and virucide

Biochemical cytocide involves the killing of pathogenic cells using biochemical, cytochemical, or other means which cause the cell to die a "natural," though perhaps highly accelerated, death. Any agent or set of conditions that stresses the metabolism or normal response mechanisms of a cell may trigger the process of apoptosis, or programmed cell death, but the level of stress is crucial.

At high stress levels, including sharp pH changes or high toxin levels, or high agitation rates, cells may die instantly by necrosis, largely because they have no time to respond to the stimulus. In necrotic cytocide, neighbouring cells are showered with cellular debris as the necrotic cell swells and lyses, an event likely to cause wider injury and a process to be avoided in nanomedical situations wherever possible.

At intermediate levels of cell stress, the cell is injured but not killed, and so has time to activate its own suicide program. The contents of the dying cell are retained within sealed vesicles until removed through phagocytosis, thus the cell dies in a controlled and tissue-friendly way. At low levels of environmental stress, cells can, for example, switch on the production of heat shock proteins which helps them to survive until the stress is removed, largely by accelerating the refolding of proteins which have been damaged by heat and some other insults. Similarly, UV radiation activates a DNA damage response, cell starvation activates the proteasome protein-recycling response, and glucocorticoids might activate responses to eliminate all nonessential functions.

Of course, once a certain stress threshold is passed and survival is deemed impossible, death by apoptosis ensues. Nanomedical biocytocidal instrumentalities should seek to dispose of cells by apoptotic methods, or by functionally equivalent techniques such as phagocytic flagging or cell division arrest.

Apoptosis

It is believed that all nucleated (eukaryotic) cells of all multicellular life on Earth incorporate an evolutionarily conserved self-destruct mechanism called programmed cell death, cell suicide, or simply apoptosis. Scattered reports on cell death have appeared in the literature (and were initially disbelieved) for more than a century, but advances in cytochemistry in the 1980s and early 1990s stimulated an explosion of interest in apoptosis, with ~2500 papers published during 1989-1994 and ~20,000 publications during 1994-1998.

Exactly what is apoptosis? Programmed cell death is the outcome of a programmed intracellular cascade of genetically determined steps. During apoptosis, the eukaryotic cell disassembles its DNA and breaks up its contents into membrane-wrapped packets which can then be cleared away without causing inflammation. Animals use apoptosis to eliminate extraneous, virus-infected, or otherwise dangerous cells. Apoptosis plays a central role both in development and in homeostasis of metazoans. Cells die by apoptosis in the developing embryo during morphogenesis or synaptogenesis, and in the adult animal during tissue turnover in the skin or gut or at the end of an immune response. Tight coupling of cell death and cell multiplication ensures in some tissues a constant, controlled flux of fresh cells which are crucial to the preservation and optimal functioning of the adult organism. Such tissues include those which are environmentally exposed, inside or outside, primarily epithelial tissue and severely insulted liver tissue, plus cells involved in reproductive function.

Apoptosis requires specialized machinery. The central component of this machinery is an irreversible proteolytic system involving a family of protein-chopping enzymes now called "caspases" from "cysteine-containing aspartate-specific proteases". Caspases are expressed as 30-50 kD proenzymes with an NH_2-terminal domain, a large ~20 kD subunit, and a small ~10 kD subunit. These enzymes participate in a cascade that is triggered in response to proapoptotic signals and culminates in cleavage of a set of proteins, ultimately resulting in an orderly disassembly of the cell.

Such complex proteolytic systems involve a combination of regulatory proteases, cofactors, feedbacks, and thresholds that converge to control the activity of an effector protease, which in turn carries out the function of the whole process. These systems keep the effector protease inactive but are able to rapidly activate large amounts of it in response to minute quantities of an appropriate inducer. Survival signals from the cell's environment and internal cellular integrity sensors normally hold a cell's apoptotic machinery in check.

Four kinds of events can trigger a suicidal cascade in a eukaryote. First, if a cell loses its normal contact with its surroundings or sustains irreparable internal damage, then that cell initiates apoptosis. Second, a cell that simultaneously receives conflicting signals driving or attenuating its division cycle also undertakes apoptosis. Third, the immune system can actively direct individual cells to self-destruct, an event called "instructive" apoptosis. Fourth, perhaps a majority of all apoptosis occurs during gestation and development, probably in response to external signal molecules or to a lack of sufficient levels of molecular signal gradients involved in biostructural development regulation.

Death receptors—cell surface receptors that transmit apoptosis signals initiated by specific death ligands—play a central role in instructive apoptosis. Death receptors can activate death caspases within seconds of ligand binding, causing the apoptotic demise of the cell within hours. Death receptors belong to the tumor necrosis factor (TNF) receptor gene superfamily. (TNF is produced mainly by activated macrophages and T cells in response to an infection.) The death receptors contain a homologous cytoplasmic sequence termed the "death domain." Death domains enable death receptors to engage the cell's apoptotic machinery, and mediate functions that are distinct from or even counteract apoptosis. In 1998, the best characterized death receptors were CD95 and CD120a, but others were known including avian CAR1, Death Receptor 3 R4, and DR5.

When a caspase cascade is triggered, the first task is to inactivate proteins that protect living cells from apoptosis. This accomplished, caspases next begin the direct disassembly of cell structure. One example is the destruction of the nuclear lamina, the rigid structure underlying the nuclear membrane that is involved in chromatin organization. Lamina is formed by head-to-tail polymers of intermediate filament proteins called lamins. During apoptosis, lamins are cleaved at a single site by caspases, causing the lamina to collapse, contributing to chromatin condensation.

Caspases also reorganize cell structures indirectly by cleaving several proteins involved in cytoskeletal regulation, including gelsolin, focal adhesion kinase (FAK), and p21-activated kinase 2 (PAK2). Cleavage of these proteins results in deregulation of their activity. Dissociation of regulatory and effector domains is another hallmark of caspase function. For example, they inactivate or deregulate proteins involved in DNA repair (such as DNA-PK_{CS}), mRNA splicing (such as U1-70K), and DNA replication (such as replication factor C).

In 1998, 13 mammalian caspases—named caspase-1 to caspase-13 —were known. Caspases cut their substrate proteins at tetrapeptide sites with high specificity and with characteristic motifs. For example, caspase-3 recognizes DEVD sequences and, during apoptosis, cleaves and inactivates several significant cellular proteins in the cytosol, nucleus and cytoskeleton. Caspases participate in apoptosis "in a manner reminiscent of a well-planned and executed military operation;" they:

1. cut off contacts with surrounding cells, so the cell balls up;
2. import calcium, strongly complexing phosphates;
3. shut down DNA replication and repair, interrupt splicing, disrupt the nuclear structure and condense the chromatin;
4. induce the loss of microvilli;
5. destroy cellular DNA by mobilizing apoptotic-unique nucleases to cleave the double helix at regular intervals in an orderly fashion, first into large fragments of 300-750 kilobases, followed by fragments of ~50 kilobases, then finally oligonucleosomal length DNA fragments of ~200 base pairs giving a characteristic pattern by gel electrophoresis called "*DNA laddering*");
6. reorganize the cytoskeleton;
7. induce the cell to display signals in the outer membrane that mark the cell for phagocytosis by neighbouring cells or macrophages.
8. induce downstream blocking of mitochondrial respiratory chain components, resulting in mitochondrial malfunction (via transcriptional induction of redox related genes, the formation of reactive oxygen species, and finally the oxidative degradation of mitochondrial components including dissipation of mitochondrial inner transmembrane potential and the release of cytochrome c through the outer mitochondrial membrane), with lysosomal involvement.
9. initiate membrane biebbing (surface blisters) and cell shrinkage; and finally
10. disassemble the cell into multiple membrane-enclosed vesicles called apoptotic bodies without destroying organelle membranes.

In vivo, this process culminates with the engulfment and phagocytization of apoptotic bodies by other cells, preventing inflammation and other complications that would result from a release of intracellular contents. These changes occur in a predictable, reproducible sequence and can be completed within 30-60 minutes.

Many different ways of activating the apoptotic process are known or suspected. Simple detachment of tissue cells from all contacts with the ECM or manipulation of cell shape have been shown to induce apoptosis experimentally. Apoptosis can also be indirectly initiated by a variety of cellular insults that can damage DNA, including ultraviolet and X-irradiation, hypoxia, and chemotherapeutic neoplastic drugs (e.g., alkylating agents such as nitrogen mustard derivatives, antimetabolites and plant alkaloids) which inflict cell damage that is then translated via the Bcl-2 protein family through several poorly understood steps into activation of caspase-9. Proteins that sense DNA damage and help trigger apoptosis also affect the cell cycle—stopping cell division so the damage can be repaired or making the decision (with the activation of tumor suppressor p53) that the damage has gone too far and the cell must die. B and T lymphocytes undergo apoptosis in response to anti-IgM antibodies and dexamethasone (a glucocorticoid), respectively. The cell death program can also be activated by inhibition of proteasome function, or by infection with a wide variety of viruses.

Caspases can be directly activated by two distinct intracellular mechanisms that interact with the death receptor complexes. First, because all caspases have similar cleavage specificity, the simplest way to activate a procaspase is to expose it to a previously activated caspase molecule. This caspase cascade is used extensively by cells for the activation of the downstream effector caspases: caspase-3, caspase-6, and caspase-7.

A second method for activating caspases is the use of apoptotic chaperones, which herd together inactive proenzymes to increase their local concentration and ease them into conformations that promote their activation. This "induced proximity" was first observed in caspase-8, an initiator caspase that acts downstream of the CD95 death receptor. Upon trimeric ligand binding, CD95 receptor molecules are aggregated into a membrane-bound complex. This signaling complex recruits, via the receptor-bound adapter protein FADD, several procaspase-8 molecules, resulting in a high local concentration of procaspase-8. Under these conditions, the low protease activity inherent to procaspases is sufficient to drive intermolecular proteolytic activation of the receptor-associated procaspase-8 molecules. Proximity-induced activation may also be used to activate caspase-9. Procaspase-9 activation involves a complex with the cofactor Apaf-1 through the CARD (Caspase Recruitment Domain), but activation of caspase-9 also requires cytochrome c (released into the cytosol by mitochondria) and deoxyadenosine triphosphate, indicating that caspase activation may require multiple cofactors.

Death receptors can be expressed on both normal and cancerous cells in the human body, so the challenge for conventional drug-based therapy is to find some way to activate death receptors selectively on cancer cells only. For medical nanorobots, such selectivity should be simple and routine using multiple chemosensors, a benefit that is characteristic of most nanorobot-based therapeutics. If caspase cascade amplification is sufficient to permit single-site activation of the cascade, then in theory an extracellular nanorobot intending cytocide could press onto the outer surface of a target cell an appropriate ligand display tool. This tool might contain suitably exposed trimeric CD95L ligand (binds to the extracellular domains of three CD95 death receptors), TNF or lymphotoxin a (binds to CD120a), Apo3L ligand aka. TWEAK (binds to DR3), or Apo2L ligand aka. TRAIL (binds to DR4 and DR5). The binding event would then activate a single death receptor complex, potentially triggering the entire irreversible cytocidal cascade. If necessary, multiple display tools could be employed. This technique avoids the storage requirement for bulky consumables onboard the medical nanorobot.

As another approach, molecular sorting rotors could be used to selectively extract from the cytoplasm specific crucial molecular species of IAPs (Inhibitors of Apoptosis) that can hold the apoptotic process in check. Examples include survivin, commonly found in human cancer cells, the transcription factor NF-kB, and Akt, which delivers a survival signal that inhibits the apoptosis induced by growth factor withdrawal in neurons, fibroblasts, and lymphoid cells. Conversely, decoy receptors (DcRs) that compete with DR4 and DR5 for binding to Apo2L could be saturated with intrinsically harmless but precisely engineered intracellular "chaff" ligands. With IAPs removed or DcRs blockaded, apoptosis may be free to proceed.

Multicellular plants and animals, human fibroblasts, and even slime molds show various forms of apoptotic cell death. Other processes of programmed cell death that may be distinct from either apoptosis or necrosis have been reported in lung fibroblasts, the caspase-free yeast *Saccharomyces cerevisiae*, and other cells. In 1998, apoptosis had not yet been reported in any bacterial species, nor was it expected because conventional evolutionary logic does not favour the emergence of cell suicide in the case of obligate single-celled organisms. Autolysis has been reported in bacteria in specialized circumstances, including addiction modules and phage exclusion, though these cases, while sometimes termed "programmed cell death," all involve a (messy) necrotic cytocidal outcome or self-damage short of cell death, rather than a (clean) apoptotic cytocidal outcome. Apoptosis also is not found in viruses.

It is theoretically possible that an artificial plasmid could be designed that embodies an engineered self-contained apoptotic system capable of unleashing a non-necrotic cell death process inside a prokaryote. A single such proapoptotic artificial plasmid (similar to artificial chromosomes, already in use by 1998) could then be injected into the target bacterium, giving the desired clean cytocidal outcome. R. Bradbury suggests the following useful components:

1. a restriction enzyme to fragment the bacterial DNA, that is effective against that specific bacteria;
2. DNase to digest the DNA entirely, if not already present;
3. RNase to digest the RNA entirely; if not already present;
4. one or more proteases such as trypsin, to digest internal proteins; and
5. one or more lipases to digest any lipids.

A lambda-phage container holding an engineered vector with only the first two capabilities could still effectively "neuter" the bacterium. Such a biorobot could be relatively harmless to eukaryotic cells, assuming its enzymes lacked nuclear localization targeting sequences; cleanup could be effectively handled by the immune system. Bradbury also suggests another approach using a DNA replication blocker such as antisense DNA polymerase, in combination with a mitotic promoter, to induce the cell to attempt to divide forever while lacking any DNA in the progeny. Essential biomolecules would eventually be diluted to such a low level that growth would radically slow and perhaps even death would result. This second approach might yield a faster decline in bacterial activity than simple neutering, due to the diversion of bacterial resources in an attempt to regain intracellular molecular balances. In either approach (as with mechanical approaches; it is necessary to avoid fragmenting the bacterium in a way which allows release of superantigen molecules such as LPS which could cause an immune system overreaction, leading to toxic shock.

Phagocytic flagging

Target cells may be flagged with biochemical substances capable of triggering a reaction by the body's natural defensive or scavenging systems. For example, novel recognition molecules are expressed on the surface of apoptotic cells. In the case of T lymphocytes, one such molecule is phosphatidylserine, a lipid that is normally restricted to the inner side of the plasma membrane but, after the induction of apoptosis, appears on the outside. Cells bearing this novel surface molecule are then capable of being recognized and removed by phagocytic cells. Seeding the outer wall of a target cell with phosphatidylserine or other molecules with similar action could activate phagocytic behaviour by macrophages which had mistakenly identified the target cell as apoptotic. Loading the target cell membrane surface with B7 costimulator molecules permits T-cell recognition, allowing an immunologic response via what has been termed .the immunological synapse. This tagging operation should work well against cells which have an apoptotic response which can be triggered by cytotoxic T cells—such as human cancer cells and cysts.

Aside from false apoptosis, removal or destruction of cell-surface MHC class I receptors might convince the immune system that the target cell was not native (e.g., "self"), making the target susceptible to attack by natural killer NK lymphocytes. (Macrophages should be able to engulf a eukaryotic cell, as they are large enough to absorb protozoa.) Alternatively, the target cell's surface would be salted with a non-self antigen-containing MHC Class I molecule, or with self MHC Class I molecules with foreign peptide attached (particularly one for which the patient has already been vaccinated), with the expectation that the cell would mistakenly be identified as foreign, and thus be phagocytosed. Other methods of engineering chemical reactivity by cell surface remodeling are readily envisioned, such as flagging with lymphokines or oligohistidine-tagged fusion proteins. But the best approach may be to seed the target cell with antigens for which the patient already has a high titer of circulating antibodies, which then will bind to the antigens and attract the immune system. Antibodies occur in higher concentrations than B/T cells and therefore would presumably trigger a faster response. As a general rule, cells with bad MHC molecules are targeted for apoptosis, while cells with antibodies attached are targeted for phagocytosis.

Cell division arrest

In some circumstances it may be useful to eliminate a target cell's ability to replicate via cell division, thus rendering it incapable of further growth. For example, many prokaryotic microorganisms respond to starvation by entering growth-free stationary phases or by forming dormant spores; nanorobot manipulation of cellular biochemical states possibly could induce stasis in bacteria, though spore formation should be avoided. The Bcl-2 protein family can modulate eukaryotic cell cycle progression; under suboptimal growth conditions, Bcl-2 promotes exit into quiescence and retards reentry into cycle. A gene fragment on human chromosome 4, called *mortality factor* 4 (morf4), when added to cancerous (eukaryotic) cells in vitro converts proliferating cells into senescent cells that have stopped dividing. Ceramide, the breakdown product of sphingomyelins, can drive a eukaryotic cell either to cycle arrest, or to apoptosis, depending upon other conditions within the cell, and p16 and p27 are known *eukaryotic cell* cycle inhibitors. Cytochalasins, which are derived from molds, interfere with the division of cytoplasm, inhibit cell movement, and can cause extrusion of the interphase nucleus.

Eukaryotic mitosis has four distinct stages (prophase, metaphase, anaphase, telophase) and interphase (between divisions), each associated with distinct biochemical and cytoarchitectural states. Thus it is not surprising that specific biochemical agents have been found that are capable of selectively halting, accelerating, or altering mitosis at any of the many stages.

Many chemotherapeutic agents function by eliminating the cell's ability to undergo mitosis. For example, plicamycin (mithramycin) is believed to form a complex with DNA which inhibits DNA-dependent or DNA-directed cellular RNA and enzymatic RNA synthesis—48-hour lethal dose in human HeLa cancer cells is ~ 1 part/million by weight or $\sim 10^5$ molecules/cell. Eukaryotic cell cycling may be interrupted by inhibitors of RNA polymerases I, II, and III (which produce RNAs specific to different functions in the cell), or by inhibitors of uracil synthesis or other steps in general RNA base production which would interfere with RNA synthesis overall. Vincristine applied at similar dosage (e.g., $\sim 10^{-18}$ kg/cell) inhibits microtubule formation in the mitotic spindle, resulting in an arrest of dividing cells at the metaphase stage. Large doses lead to necrotic cytocide; a smaller dose may trigger apoptosis.

A theoretical enzyme capable of selectively lysing DNA telomeres in the nucleus would prevent cell replication, but would probably trigger apoptosis due to massive chromatin damage because TTAGGG sequences are scattered throughout the chromosome. Telomere length affects eukaryotic cell division only indirectly, either because short telomeres activate a DNA damage response that blocks cell division or because genes near the ends of chromosomes which are essential for cell division may be lost or misregulated. Telomerolytic enzymes would be useless against common bacteria (e.g., *E. coli*) and viruses (e.g., SV40) which have circular DNA chromosomes, or viruses such as the adenoviruses that cause bronchitis and pneumonia which have linear DNA locked with a terminal protein—but no telomeres in either case. Alternatively, R. Bradbury suggests the introduction of DNA (or RNA) for antisense sequences, for genes essential for cell cycle progression or for proteases which cut or digest these same proteins, as for instance an antisense gene for cyclin or one of the many cdc genes—if phosphorylation or dephosphorylation of the cyclins or cdc genes is occurring, the introduction of a phosphatase or kinase could easily block the entire cell division process.

Chemical poisoning

There are a great many (often inefficient) ways to poison a cell or virion simply by releasing chemical agents near, on, or within it, but if done incautiously this approach may lead to necrotic, rather than apoptotic, cell death. Here are a few examples of some well-known classes of biocidal agents, many of which may be inappropriate for use by in vivo medical nanorobots:

Phagosomal biochemicals

Both types of phagocytic cells (e.g., neutrophils and macrophages) contain specialized organelles that fuse with newly formed phagocytic vesicles (phagosomes), exposing phagocytosed microorganisms to a barrage of enzymatically produced, highly reactive molecules of superoxide (O_2^-) and hypochlorite (HOCl, the active ingredient in bleach), called the "oxidative burst" that punctures cell walls, and to a concentrated mixture of lysosomal hydrolases. Human neutrophils also use antimicrobial peptides such as the serprocidins (e.g., proteinase 3, azurocidin, and cathepsin G, a metabolic inhibitor) against fungi and bacteria.

Cytolytic enzymes

Lysozymes, present in tears, nasal mucus and sputum, destroy the cell walls of many airborne Gram-positive bacteria by catalyzing the hydrolysis of b-1,4 linkages between N-acetyl muramic acid and N-acetyl glucosamine (peptidoglycan degradation) in bacterial cell walls—causing the bacteria to burst open, spilling their contents. Lysozyme, zymolase, glucalase and lyticase are frequently used with bacteria and yeast cells to dissolve coats, capsules, or capsids. Granulysin released by cytolytic T lymphocytes directly kills intracellular *Mycobacterium tuberculosis* by altering the membrane integrity of the bacillus.

Organelle poisons

Certain drugs and other substances, if injected into the cytoplasm, may interfere with the workings of specific organelles or cell subsystems. For example, the fungal metabolite drug brefeldin A disrupts the Golgi, causing it to collapse back into the ER. Colchicine, vinblastine and vincristine bind to tubulin, causing microtubules to disassemble, griseofulvin (an antifungal agent) prevents microtubule assembly, and taxol inhibits microtubule depolymerization, while cytochalasin B inhibits the polymerization of cytoskeletal actin microfilaments. Adociasulfate-2 is a kinesin motor inhibitor. High concentrations of vitamin A weaken the lysosomal membrane. Proteasome inhibitors such as vinyl sulphone are usually lethal for the eukaryotic cell. Oligomycin inhibits the F_0F_1-adenosine triphosphatase (ATPase) proton pump of the mitochondrial inner membrane, and cyanide poisons mitochondrial oxidative phosphorylation. Nitric oxide (cone. ~60 nanomolar) inhibits mitochondrial respiration, specifically by inhibiting cytochrome oxidase by outcompeting oxygen at the oxygen binding site; NO also prevents viral replication by inactivating a crucial protease. Antipsychotic drugs such as chlorpromazine, thioridazine, and fluphenazine are potent peroxisome inhibitors. The nucleus of an oocyte is ejected from a cell treated with etoposide and cycloheximide (chemical enucleation), microtubule poisons such as colchicine, colcemid and vinblastine cause extrusion of nuclei, and EDDF is involved in erythroid cell denucleation; there is at least one report of nuclear extrusion in lymphocytes. Selected organelle populations (e.g., mitochondria) could also be ubiquinitized.

Antibiotics

Drugs that prevent bacteria from multiplying (bacteriostatic) or that kill bacteria outright (bactericidal) are called *antibiotics*; in 1998, ~100 such drugs were PDA approved for U.S. use. Major groups include the aminoglycosides, cephalosporins, macrolides, penicillins, polypeptides, quinolones, sulfonamides, and tetracyclines; many are naturally-derived products. All act by interfering with protein synthesis, cell wall construction, or DNA replication. For example, *gonococci* isolated in the preantibiotic (pre-resistance) era were inhibited by benzylpenicillin ($C_{16}H_{18}N_2O_4$ S, MW = 334 daltons) in concentrations as low as ~7 molecules/micron^3; in the 1990s, some highly resistant bacterial isolates required ~1000-fold higher concentrations. Vancomycin derivatives are active against Gram-positive bacteria at 10-100 molecules/micron^3 Antibiotics are available with protein/RNA synthesis inhibition activity against either prokaryotic or eukaryotic cells.

Bactericidal phages

Bacteriophages (viruses that infect bacteria) are capable of penetrating bacterial membranes and delivering foreign DNA which can take control of all metabolic processes; in ~700 sec after penetration, the first complete virion particles begin appearing in the cytoplasm; at ~1500 sec, the bacterium bursts, necrotically liberating ~200 virus particles. These viruses are often very host-specific. Still, it may be possible to design artificial general-purpose bacteriophages that are capable of disabling or destroying all bacterial DNA, or are capable of replicating more bacteriophage particles to which only the targeted bacterial cells are susceptible. Engineered macrophages or artificial neutrophils might also be deployed to achieve targeted bacterial digestion. Reaching intracellular parasites such as mycoplasmas and rickettsias might be one of the biggest challenges for such biorobots.

Bacteriocins

Probably 99% of all bacteria generate at least one bacteriocin, small proteins that function as narrow-spectrum antibiotics that may have developed as poisons to kill competing bacteria. In 1998, ~80 bacteriocins were known, most of them produced by fermentation microbes, including nisin which is used in 45 countries including the U.S. as a commercial food additive for pasteurized egg products. Bacteriocins seem to work by entering the outer membrane of a susceptible bacterium, congregating in groups, and forming pores that allow the unregulated outflow of essential ions. The target bacterium begins breaking down ATP in a vain attempt to produce enough new protons to recharge the membrane, a futile cycle that quickly results in exhaustion of bacterial ATP. Bacteriocins are most effective in acid and least effective in salty environments. Bacteriocins of Gram-positive bacteria such as *Listeria* and *Clostridium botulinum* are ineffective against Gram-negative bacteria such as *E. coli* and *Salmonella* which have a protective double-walled outer membrane. In 1998, several Gram-negative bacteriocins were known. One example is the colicins, water-soluble cytotoxins secreted by and active against *E. coli*, that form voltage-sensitive ion channels in the bacterial inner membrane that kill the cell by selectively siphoning out key cell nutrients and inhibit protein synthesis.

Porins and superporins

C. Sublette notes that it should be possible to design and insert into a cell a piece of RNA or DNA designed to produce high levels of a "superporin" protein that migrates to the cell membrane, then self-assembles into a large channel in the membrane several orders of magnitude larger than an ion channel, allowing a large, lethal ion influx or outflux. For example, normal cytosolic Ca^{++} concentration ranges from 60-3000 ions/micron^3; blood plasma concentration is ~10^6

ions/micron^3, but even cytosolic concentrations as low as ~10^4 ions/micron^3 may be toxic. Possible templates for such subunits are the bacterial porins, minus their charged amino acid residues lining the inner passage which produce nonspecific aqueous diffusion channels across the outermost LPS bacterial membrane, and granulysin, perforin and members of the amoeba pore family which are thought to damage target cell membranes by inducing formation of microscopic pores.

Ionophores

Ionophores are small hydrophobic molecules that dissolve in lipid bilayers, thus increasing ion permeability of cell membranes. Most are synthesized by microorganisms, presumably as biological weapons to weaken their competitors. There are two classes of ionophores—(highly temperature-sensitive) mobile ion carriers, and channel formers—both operating by shielding the charge of the transported ion so that it can penetrate the hydrophobic interior of the lipid bilayer. (Ionophores permit net movement only down their electrochemical gradients, since no energy sources are available.) Valinomycin, a ring-shaped polymer that increases K^+ permeability of membranes, is an example of a mobile ion carrier. Another example is A23187, which acts as an ion-exchange shuttle, carrying two H^+ out of the cell for every divalent cation (such as Ca^{++} and Mg^{++}) carried in. Transport rates through a mobile carrier are ~2×10^4 ions/ sec. A channel-forming ionophore is gramicidin ($C_{148}H_{210}N_{30}O_{26}$, MW = 2822 daltons), a 15-residue linear polypeptide with all hydrophobic side chains. Two such molecules come together in the bilayer to form a transmembrane channel that selectively allows monovalent cations (most readily H^+, K^+ somewhat less readily, Na^+ still less readily) to flow down their electrochemical gradients at a rate of ~2×10^7 cations/sec. The dimers are unstable, constantly forming and dissociating, with average channel open time ~1 sec. Gramicidin is produced by *Bacillus brevis* and is active against Gram-positive cocci and bacilli. A 5 microgram dose kills, in vitro, 10^9 *pneumococci* or group A *streptococci* in 2 hours at 30°C, giving a fatal dose of ~1000 molecules/micron^3 assuming a ~1 cm^3 culture volume in this experiment.

Channel blockers

Cells import and export nutrients, ions, water, and wastes through a variety of gated channels and transporter molecular pumps. If these channels or pumps are permanently blocked, cytocide may ensue. Blocking the ~10^3/micron cellular transport systems embedded in a typical eukaryotic cellular membrane requires ~10^3/micron "steric plugs" that permanently jam in the throat of critical ion channels or pumps. For example, the acetylcholine receptor channel narrows to ~0.65 nm at its waist, so an efficiently designed steric plug could have a volume of ~1 nm^3 or ~10^{-21} gm. (By comparison, a single molecule of 318-dalton tetrodotoxin, or puffer fish toxin, has mass ~0.53×10^{-21} gm; apparently, saxitoxin (shellfish toxin, human IV lethal dose ~68 micrograms) and palytoxin are slightly smaller, whereas botulin toxin is a fairly high-mass zinc metalloprotease protein that inhibits release of acetylcholine at the neuromuscular junction.) Plugging all channels in the surface of a $(2 \text{ micron})^3$ bacterial cell (~10 micron) or a $(20 \text{ micron})^3$ tissue cell (~1000 micron) requires ~10^4 or ~10^6 plugs, respectively, a nanorobot-dispensed ~10^{-5} micron^3 or ~10^{-3} micron^3 dosage of plugs at the target cell, assuming ~100% efficiency. A mean migration distance per plug of ~1 micron gives a Brownian diffusion time of ~1 millisec, so a micron-scale nanorobot with 0.1 micron^3 of onboard storage carries sufficient plug-doses to incapacitate ~10^4 bacterial cells in ~10 sec or ~10^2 tissue cells in ~0.1 sec, ignoring cell-to-cell travel time. Assuming a mean interbacterial separation of ~100 microns (an implied pathogen number density of ~$10^6/cm^3$ in the tissue) and a maximum nanodevice intercellular travel speed of ~100 micron/sec, a nanorobot spends ~1 sec in transit between neighbouring bacterial cells, ~0.002 sec in cell-type recognition, and ~0.001 sec killing the bacterium once the nanorobot arrives.

Channel destroyers

Some smaller number (perhaps as few as one) of an enzyme-like toxin molecule capable of destroying either the steric or the electrochemical specificity of each ion channel by permanently altering the physical structure of that channel. For example, in the case of the acetylcholine receptor channel, an artificial enzyme possibly could be designed to alter the ring of negatively charged residues at the entrance to the receptor, allowing anions to enter the pore and ultimately depolarizing the cell. Enzymes typically operate at ~1000 Hz and the diffusion time across ~20 nm between adjacent ion channels is ~0.5 microsec, so a single enzyme neutralizes all 10^4 channels in a bacterial cell in ~10 sec and ~1000 artificial enzymatic molecules would incapacitate the cell in ~10 millisec.

Complement

The complement system is a group of ~20 soluble serum-resident proteins acting in combination with specific cytolysin antibodies in amplification cascades along two separate pathways to initiate reactions to foreign antigens. Complement proteins are enzymes that act on foreign cells by punching holes in their membranes by inserting lipid-soluble pores (the membrane attack complex, C5-C9), causing necrotic osmotic cytolysis. In addition, complement plus antibody designates cells to be engulfed by phagocytic cells.

Animal venoms and toxins

The bufagin from the *Bufo arenarum* toad of Argentina (arenobufagin, $C_{25}H_{34}O_6$, MW = 430 daltons) has a mean lethal toxicity of 92 microgram/kg, the equivalent of ~10^6 bufagin molecules per 20-micron tissue cell (~100 molecules/micron^3). Most reptile venoms have lethal toxicities of 200-1000 micrograms per kg of body weight. Sea anemone granulitoxin (4958 daltons) has a mouse LD50 of 400 micrograms/kg (~49 molecules/micron^3); another sea anemone protein extract has a mouse LD50 of 40 micrograms/kg. Maurotoxin, a 34-residue scorpion venom (*Scorpio maurus*), has an intracerebroventricular mouse LD50 of ~80 nanograms, or ~0.4 molecules/micron^3. The LD50 value to mice of *Vespa luctuosa* hornet venom is 1600 micrograms/kg, the most lethal known wasp venom. Venoms and toxins may not be cytocidal against all cell types—some act by interfering with nerve transmission or by paralyzing muscles.

Plant and microbial toxins

A variety of microbes manufacture neurotoxins or neurolysins that destroy the ability of human ganglion and cortical neural cells to function. A single molecule of some neurotoxins can incapacitate a cell, though the rate of action tends to be very slow. For example, botulin toxin is the second deadliest poison known (e.g., median human lethality ~5-50 nanograms/kg). The toxin acts at cholinergic nerve terminals to prevent acetylcholine release, a permanent physiologic denervation that causes muscle paralysis, often resulting in death by respiratory system failure in 2-8 days.

Recovery requires the sprouting of new axon twigs and the formation of new myoneural junctions, which is why the recovery time for sublethal botulin poisoning is very long, several months to a year or more. Botulin is a ~150,000-dalton protein comprised of 1285 amino acid, which implies a lethal concentration of 0.2-2 molecules per 20-micron tissue cell. Other toxins are capable of killing a cell outright with just a few molecules, such as ricin: mouse and rabbit IV LD50 ~400-4000 nanograms/kg or 30-300 molecules per 20-micron tissue cell, possibly due to lower specificity in binding to cells and other pharmacokinetic effects. Ricin is a phytotoxic enzymatic 65,750-dalton protein taken from the seeds of the castor oil plant that cleaves the ribosome complexes, shutting down protein manufacturing. This damage is normally irreversible, killing the organism in ~4 days at minimum dosage and ~8 hours at the highest dosages, so a reasonable lower limit for killing an individual cell is probably ~10^4 sec.

Antifungals

Antifungal agents such as amphotericin B or fluconazole are fungistatic. For example, the growth of *Candida* and many other fungi is inhibited by amphotericin B ($C_{47}H_{73}NO_{17}$, MW = 924 daltons) concentrations of 30-1000 micrograms/cm^3 in vitro. A number of natural antimicrobial peptides are induced in epithelial cells at sites of inflammation; for example, lingual antimicrobial peptide (LAP), a member of the b-defensin class, has been isolated from bovine tongue and exhibits both antifungal (e.g., *Candida*) and antibacterial (e.g., *E. coli*) activity at concentrations as low as ~16 microgram/cm^3. In 1998, antifungals were known to target membrane function, metabolism, cell wall synthesis, protein and ergosterol synthesis, nuclear division, and nucleic acid synthesis and function. In those circumstances where fungal cells have learned to communicate to effect mating or cooperative behaviours, such communication may be disrupted to their detriment. If the genetic pathways involved in these behaviours include primitive forms of apoptosis, these could be exploited as well.

Antivirals

Antiviral drugs may work by interfering with viral replication processes, including cell attachment, cell uptake, viral coat removal, and viral DNA or RNA replication by the cell, and are primarily virostatic. For example, acyclovir (MW = 225 daltons) in concentrations of 0.01-13.5 micrograms/cm^3 inhibits by 50% the growth of *herpes simplex* virus in vitro. In 1998, ~20 antivirals were PDA approved for U.S. use. Antibodies may also serve a virostatic function. For example, the rabies-like vesicular stomatitis virus (VSV) rhabdovirus has a surface envelope with ~1200 identical glycoprotein molecules that form a regular and densely ordered pattern of spike tips. Rhabdoviruses are neutralized if they cannot dock with their cellular receptors; this requires a minimum of 200-500 IgG antibody molecules bound per virion. Another challenge for medical nanorobots will be the retroviruses, which insert their genome into the host DNA, thus will require chromosomal editing or replacement to remove.

Iodine-based microbicides (iodophors)

In tincture of iodine, all of the iodine is in the free form at ~10% by weight and is readily available for instantaneous reaction and killing of bacteria in a few seconds. The minimum lethal iodine dose for a 2-micron bacterium (assuming complete exhaustion of a 0.1-micron perimicrobial iodophor layer) may be of order ~0.001 picomoles or ~0.1 picogram iodine (~10^8 molecules), which is ~3% of the mass of a bacterium—not terribly efficient, as expected for a broad-spectrum agent. Iodoacetate inactivates most cytoplasmic enzymes and blocks anaerobic metabolism, or glycolysis (fluoride is also a glycolytic poison). Yodoxin (64% organically bound iodine) is amoebicidal. The broadest-known spectrum iodine-based microbicide is betadine (povidone-iodine) 10% solution, employed in surgical scrub and many other general antiseptic uses. Betadine kills most bacteria in 15-30 sec, and also kills viruses, fungi, yeasts and protozoa; no bacterial resistance has been reported.

Silver

Silver foil has been used in bacteriocidal wound dressings, and silver metal has been employed for centuries to purify water. In 1998, there were more than 600 silver-based antibacterial products available in Japan including silver-impregnated pens, floppy disks, calculators, ATM machines, floor tiles, plastic food wrap, socks, shirts, public park sand, and toilet seats. At least 20 companies in the U.S. offer silver/copper-based systems for swimming pool sanitation and commercial air conditioning cooling towers. The passive dissociation of silver from the metallic phase into a wound with antimicrobial results and the antiseptic action of silver compounds are well known. A quantitative study of the metal's antimicrobial properties in vitro found that a silver ion concentration of ~25 micrograms/cm^3 (~10^5 ions/micron^3)) reduced the number density of *Staphylococcus* and *Pseudomonas* bacterial cells and *Candida* fungal cells by a factor of 1-10 million relative to microbial control aliquots.

Platinum, bismuth, and other elements

Platinum-based cisplatin ($PtCl_2H_6 N_2$, MW = 300 daltons) is an antitumor and bacteriostatic drug that interacts with tumor-cell DNA by forming a Pt-GG intrastrand crosslink as the critical lesion leading to cytotoxicity. A typical ~100 mg whole-body dose produces a mean concentration of ~2000 molecules/micron^3. Colloidal bismuth subcitrate exerts a direct antimicrobial effect against *H. pylori* (gastritis) at concentrations of 4-25 micrograms/cm^3 (6000-38,000 molecules/micron^3), with treated bacteria showing deposits of bismuth on their surface and internally. Mercury (chronic toxicity in blood ~600 atoms/micron^3), lead (toxic in blood at ~2900 atoms/micron^3), and arsenic (toxic in blood at ~4800 atoms/micron^3) are general poisons that have been used for medicinal purposes. Indirect poisons such as iron and copper catalyze free-radical formation. Interestingly, bacterial growth is constrained by essential nutrients, and human immune cells scavenge and sequester iron when faced with a bacterial threat; nanorobotic "deferritization" of a bacterium could kill the organism.

Sugar

Medical lore has it that sugar, and especially honey, makes an excellent antibacterial disinfectant. Blood glucose levels of 0.01-0.07 M in untreated diabetics cause tissue and cellular damage, at least partly via osmotic cellular dehydration (honey and NaCl salt have a similar effect), and protein glycosylation and Maillard reaction which diminish enzyme function; ~15% cytosolic glucose concentrations are toxic.

Hydrogen-bond and disulfide-bond breaking agents

At high concentrations, guanidine hydrochloride disintegrates the bacterial Slayer coat which is held together only by noncovalent forces. Local heating (hyperthermia; also disrupts H-bonds. Other agents are available to break disulfide bonds, disrupting many classes of proteins.

Nuclear alkalination and other nucleic disruption

RNA breaks down in slightly alkaline environments (e.g., pH > ~ 8.0; to increase the pH of a typical cellular nucleus from 7.2 to 8.0 would require the injection of ~10^5 OH^- ions. RNA is also broken down by RNases; DNA is broken down by DNases, or by using restriction enzymes.

Tissue liquefaction

A 2-hour exposure to 0.25% trypsin · yme solution digests and breaks up the extracellular matrix, turning an organized tissue into a jumble of individual cells.

Other poisons

Cyanide causes cells to swell and lyse at concentrations of ~0.002 M (~1 × 10^6 ions/micron^3). Assuming serum reference levels, ethanol should be cytotoxic at ~0.09 M (~50 × 10^6 molecules/micron^3). Ethylene oxide (C_2H_4O) gas is a microbial sterilant and fumigant; chlorine dioxide (chlorite ion) is an antimicrobial sometimes found in mouthwashes. Chlorine and bromine are used as antibacterials in swimming pools and hot tubs. Most chemotherapy agents also kill cells. Common lab detergents such as Triton X-100 and tributyl phosphate readily destroy bacteria and the majority of viruses sheathed in protein envelopes.

Nanorobot biocidal delivery vehicles can be used to present the chemical agent on a cell-by-cell basis. For example, a major whole-body infection involving $\sim 10^{12}$ pathogens (e.g., $\sim 10^6$ bacteria/cm^3) with each microbe capable of being killed by $\sim 10^5$ precisely-delivered biocidal molecules each of molecular weight ~ 1000 daltons requires a minimum treatment dose of $\sim 10^{17}$ biocidal molecules or ~ 0.1 mm^3 of material. This dose could be carried and dispensed in ~ 1000 sec by one billion ~ 1 micron^3 "pharmacyte" nanorobots (total volume of therapeutic nanorobots ~ 0.001 cm^3) assuming each nanorobot has ~ 0.1 micron^3 of internal storage for the biocidal agent and assuming as before a mean ~ 1 sec cell-to-cell transit time.

Mechanical Cytocide and Virucide

Mechanical cytocide involves the killing of biological cells or the destruction of virion particles partially or primarily by mechanical, rather than by strictly biochemical, means. Such means may include disablement, partial physical destruction of critical components, or even complete disassembly of the target cell, bacterium, or viral particle.

Digest discharge (DD)

After virion acquisition as previously described, the particle is placed in a leakproof transfer chamber. The virus is then pistoned into a morcellation chamber where it is chopped by orthogonally traveling diamondoid blades into ~ 10 nm pieces, greatly increasing the reactive surface area of the virus material. The morcellate is then pistoned into a reaction chamber. Capsid-specific proteinases and peptidases are introduced, reducing all virus proteins to amino acids, which are removed from solution by molecular sorting rotors and discharged. Lipases are required for digesting enveloped viruses. Deoxyribonuclease and ribonuclease enzymes are also introduced, reducing viral DNA or RNA to nucleotides which are themselves removed from the solution by a second set of molecular sorting rotors and discharged. The enzymes remaining in solution are pumped back into storage vessels via sorting rotors, and the reaction chamber is ready for the next cycle.

For example, a 70-nm adenovirus particle containing 8% DNA by weight is reduced to 1.7×10^{-16} gm ($\sim 10^6$ molecules) of amino acids and 1.4 × 10^{-17} gm (~30,000 molecules) of nucleotides. Natural bloodstream concentrations of all amino acids are typically ~5 × 10^{-5} gm/cm^3 (~3 × 10^{11} adenovirus equivalents per cm^3), and ~10^{-6} gm/cm^3 (~7 × 10^{10} adenovirus equivalents per cm^3) for nucleotides, so the maximum "safe." viral material discharge rate appears nucleotide-limited to ~10^{11} virus particles per cm^3 which is at or above the typical virion particle density in serious infections. Hence, digested-virus material discharges should not significantly augment natural serum concentrations of amino acids or nucleotides.

Energy costs are minimal. Single-virus mincing costs ~0.1 pW of power for a duration of ~8 millisec. Representative enzyme operating frequencies (turnover numbers) range from 1-2 KHz for lactate dehydrogenase and penicillinase to 15-100 Hz for DNA polymerase and chymotrypsin, giving digestion times on the order of 1-100 millisec when an excess of enzyme is present. Out-rotoring all final-product amino acids costs at most ~ 20 pW for at most ~10 millisec duration using ~100 sorting rotors per each of the 20 essential amino acid types. Pumping a generous ~100 virus-volumes of enzyme-rich solution through a 1-micron length of 60-nm diameter pipe at 1 atm pressure costs 2 pW for ~ 1 millisec duration. Hence virion processing time is of order ~100 millisec and requires at most ~30 pW during processing.

Mechanical bacteriocide

Bacteria are unicellular microorganisms capable of independent metabolism, growth, and replication. Their shapes are generally spherical or ovoid (cocci), cylindrical or rodlike (bacilli), and curved-rod, spiral or comma-like (spirilla). Bacilli may remain associated after cell division and form colonies configured like strings of sausages. Bacteria range in size from 0.2-2 microns in width or diameter, and from 1-10 microns in length for the nonspherical species; the largest known bacterium is *Thiomargarita namibiensis*, with spheroidal diameters from 100-750 microns. Spherical bacteria as small as 50-500 nm in diameter have been reported, but it has been theorized that the smallest possible cell size into which the minimum essential molecular machinery can be contained within a membrane is a diameter of ~40-50 nm. Many spherical bacteria are ~1 micron in diameter; an average rod or short spiral cell is ~1 micron wide and 3-5 microns long. Each bacterial cell consists of a mass of protoplasm enclosed within the usual thin lipid bilayer plasma membrane. Most Gram-positive bacteria are surrounded by a thick, mechanically strong but porous peptidoglycan cell wall. Gram-negative bacteria like *E. coli* surround themselves with an additional two-layer coat atop the peptidoglycan layer. This coat has an unsaturated inner lipid layer but a more rigid outer leaflet composed of an unusual lipid, called lipopolysaccharide (LPS), in which fatty acid chains are all saturated and 6-7 chains are covalently linked in a single LPS molecule.

Bacterial cells have no internal membranous surfaces and no internal organelles, although some functional compartmentalization does exist. Bacteria possess small internal vacuoles, ribosomes, and granules of stored food, and usually one or more externally-attached flagella. Some rod-shaped bacteria can form tiny spherical or oval endospores that survive when conditions become inhospitable for metabolism, such as extreme heat ($>100°C$) or desiccation (for up to 60 years). As prokaryotes, bacteria have no distinct nucleus. However, a single circular chromosome is organized into one or more compact aggregates, called ***nucleoids***, that may occupy about one-third of cell volume. *E. coli*, a well-studied cylindrical bacterium measuring 0.65 microns wide and 1.7 microns long (cell volume $\sim$ 0.6 micron^3), has one double-stranded DNA chromosome $\sim$1.3 mm ($\sim$4.2 megabases or $\sim 10^9$ daltons) in length (strand volume $\sim$0.002 micron^3), organized in $\sim$40 kilobase loops. Chromosomes spilled from lysed bacteria show $\sim$80% DNA content, suggesting loop stabilization by proteins and at least a primitive membrane-associated cytoskeleton.

There are at least three methods for eliminating an unwelcome microbe from the human body. An additional serious problem in bacterial infections is that many microbes release toxins into the bloodstream that can wreak havoc even if all the bacteria are killed. Toxin cleanup is a necessary component of any comprehensive antibacterial treatment strategy.

15

SCOPE OF NANOTECHNOLOGY

Nanotechnology is often described as an emerging technology—one that not only holds promises for society but is capable of revolutionizing our approaches to common problems. Nanotechnology is not a completely new field; however it is only recently that discoveries have advanced so far as to warrant examination of their impact upon the world around us. Nobel laureate Richard P. Feynman, in his famous speech to the American Physicist Society in 1959, said "in the year 2000, when they look back at this age, they will wonder why it was not until the year 1960 that anybody began seriously to move in this direction."

PRODUCTS OF TODAY

The value of nanomaterials in many technology areas is high because of their versatile properties. Investment in nanotechnology in the United States government has had a very steady growth; in 2004 we are approaching the $1 billion a year in investment from many different federal agencies, said Olden. Industrial investment in this area is also growing steadily. It is said that the future of nanotechnology is potentially boundless if we can avoid the pitfalls. Some of the items that exist today were a topic of science fiction a decade ago and have the potential to transform our society very quickly, said Douglas Mulhall, author of *Our Molecular Future; How Nanotechnology, Robotics, Genetics, and Artificial Intelligence Will Transform Our World.* Today, some of the nanomaterials are already being used in commerce. For example. Proctor and Gamble is using titanium dioxide (TiO_2) nanoparticles to provide transparency in sunscreen lotions, as a less toxic alternative to the organic molecules currently used as UV absorbers in many sunscreen formulations. Other nanomaterials are incorporated in sporting equipment, clothing, telecommunication infrastructure, and fuel cells.

The future of nanotechnology is potentially boundless if we can avoid the pitfalls. Some of the items that exist today were a topic of science fiction a decade ago. Other examples of current nanotechnologies are three-dimensional printing with nanoparticles, high speed computing driven by nanotechnology, and computer-assisted design software with generic algorithms that learn by themselves. Three-dimensional printers were invented in the 1990s and commercialized in 2000 and are already being used to make products such as surgical models. Surgical models have the potential to match the exact needs of an individual patient while allowing surgeons to eliminate the need for exploratory surgery when they are doing reconstructive work.

Today it is technologically feasible to manufacture programmable nano-filters that eliminate the pressure requirement for desalinization and reduce the expenses by 99 percent, meaning the end of water shortages in arid regions. It could also potentially alter regional environments and convert deserts to forests by removing large amounts of fresh water from the oceans, postulated Mulhall. With the invention of solar paint and principal solar cells based on nanoengineered organic plastics, fossil fuel dependence may begin to recede. Another nanotechnology-based invention is extremely strong materials called nanostructured *aerogels*. These materials are inexpensive and they could make structures resistant to earthquakes and hurricanes, which will have significant societal benefits, said Mulhall.

Environmental Applications for Nanotechnology

Some of the greatest potential uses or applications for nanotechnology in the environment are sensors, treatment, remediation, and green nanotech manufacturing and engineering, stated Barbara Karn of the Environmental Protection Agency. These applications can be further categorized as either reactive to existing environmental problems or proactive in anticipating and preventing future problems.

Reactive Applications

Reactive applications involve sensors, treatment, and remediation. Diverse research is being conducted on the use of nanoparticles as sensors to allow for accurate, real-time simultaneous sensing of a variety of compounds, often at extremely low concentrations and in hostile environments. Karn discussed a number of research projects sponsored by the Environmental Protection Agency that are aimed at developing applications for nanotechnology.

The research of Nongjian Tao at Arizona State University focuses on "lab on a chip" nanotechnology for on-site detection of ultratrace levels of heavy metal ions, including radioactive compounds, using a silicon chip which contains nanoscale electrodes separated by an atomic-scale gap. Thus, with the electrode position of only a few metal ions in this gap, the circuit can be completed, indicating the presence of various compounds.

Nanoscale sensors are also being investigated for detection of biological compounds such as algal toxins in the marine environment or mycobacteria present in drinking water. Robert Gawley of the University of Arkansas has developed fluorescent dendrimers displaying spatially resolved microdomains on polymer beads for the detection of different algal toxins. The binding of different toxins results in specific fluorescence wavelengths, depending upon the spatial resolution of the dendrimers on the polymer beads, which correlate to known toxins. This technology would be a less costly and time consuming alternative to current methods used to monitor shellfish populations, which are especially sensitive to marine toxins such as those responsible for large fish kills on the Eastern seaboard.

Advances in nanodetection would allow not only for detection of microbial pathogens in drinking water, but also the quantization of these organisms, noted Karn. A piezoelectric microcantilever sensor is a method of detection and quantization being developed by Wan Shih of Drexel University, which involves the use of micron-scale cantilever arms coated with specific DNA strands. Test water is run through the piezoelectric cantilever sensor and when target molecules bind the cantilever becomes heavier, changing its rate of vibration and creating an electric current. Thus movement of the cantilever by just a few nanometers correlates with the number of molecules captured on its surface, allowing for detection and quantization of specific pathogens in the water.

Treatment and remediation are two other reactive applications made possible by nanotechnology research. Nanoscale molecules used in treatment and remediation have the ability to access areas that larger molecules cannot, and can be coated to prevent reactivity with surrounding soil particles. In addition, the researchers are following the paths of these nanoparticles, so the fate and transport of these molecules as well as the clean-up efficiency is being studied, added

Karn. For example, iron oxide particles encapsulated in a protein shell can be used for the reduction of heavy metals such as hexavalent chromium, a frequent groundwater contaminant. Karn pointed out that clean-up is where nanoscale makes an impact: for example, the smaller the radius of zero-valent iron particles, the more milligrams of trichloroethylene (TCE) are reduced per milligram Fe per hour. Iron particles with a radius of 1 μm can reduce 0.186 mg TCE/mg Fe/h. However, this same type of particle with a radius of 1 nm can reduce 186 mg TCE/mg Fe/h. The nanoscale particle is three orders of magnitude smaller, but its rate of TCE reduction is three orders of magnitude greater than that of the microscale particle.

Nanotechnology will make possible great advances in our ability to retroactively solve environmental issues. Research varies in the approaches taken to reach the endpoint of treatment or remediation, said Karn. Some researchers attempt to surface-coat molecules to pull them together, and maybe detoxify them while others coat them so that the desired nanoproduct is isolated and collectable. Regardless of the strategy utilized, Karn proposes that nanotechnology will make possible great advances in our ability to clean up the environment.

Environmental Proactive Applications

Proactive applications mainly include green manufacture and green energy productions. Green energy can be produced using nanotechnology to create solar and fuel cells as potential sources of commercially available alternative clean energy sources. Green manufacturing has two aspects: the use of nanotechnology in the design process to eliminate polluting waste products at their source and, alternately, the efficient production of nanomaterials themselves.

Green manufacturing improves catalysis specificity, producing more of the desired compounds and less waste and pollution. Researchers are exploring how to stabilize nanoparticles without harmful additives that would pollute water and soil. Often these processes require less time, sometimes hours instead of days, and can potentially significantly reduce the cost of making nanoparticles. Consumption of energy during the manufacturing processes can also be reduced, and Karn noted that many aspects of nanotechnology and its application in manufacturing are devoted to the research and production of cleaner and more efficient energy sources.

Green energy can be produced using nanotechnology to create solar and fuel cells as potential sources of commercially available alternative clean energy sources.

Nanotechnology has a great future for providing solutions in the area of green energy applications, said Karn. New research is exploring the use of electrochemistry, nanoscale photosynthesis, micro fluidic biofuel cells, and photo electrochemical cells, among other endeavors, to address the growing need for safe, inexpensive, and renewable energy resources. For example, solid state lighting can be enhanced by the use of nanocrystals/polymer composites for light-emitting diodes (LED). This allows for brighter, more efficient, and less costly lighting for use in such equipment as traffic signals. Karn stated that use of LED technology could reduce the consumption of light energy up to 50 percent by 2020 in the United States alone. Reduction of carbon emissions by an estimated 28 million metric tons per year has obvious climate change and materialization implications. Economically, this translates to a savings of $100 billion in fewer than 20 years.

The challenges facing researchers with regard to fuels in the realm of transportation are three-fold: how to produce the fuel, how to transport it to the location where it is needed, and how to store the fuel safely. These questions are especially pertinent to hydrogen fuels and were addressed in President Bush's State of the Union address. Aimed at putting an appreciable number of motor vehicles on the road that are powered by hydrogen fuel by 2020, the initiative allocates $1.2 billion to research and production of these vehicles.

Nanotechnology may offer solutions in all these areas of concern, noted Karn. Nanomaterials may be used to create harder alloys and ceramics for cutting tools to increase the efficiency of manufacture. Motors may be made more efficiently by incorporating low-loss, high-performance magnets. The use of lightweight materials with low failure rates and surface tailoring of parts to produce less friction and increase resistance to wear, would decrease fuel consumption while improving the safety of ground and air transportation. Finally, the use of nano catalysts will lead to cleaner, less costly, and environmentally friendly petroleum refining and the production of more efficient and productive batteries and fuel cells, said Karn.

Sustainability is a top consideration in the quest for new fuel sources, particularly with reference to the issue of global climate change. The harm to human health and the environment that the transportation systems produce is well known; thus, with the better catalysts, better kinds of tires, better, lighter materials that nano-technologies can offer, this can help affect climate change by changing the kind and amounts of resources consumers use, said Karn. Lighter-weight materials lower the amount of fuel needed for transportation, more efficient electronics use less electrical energy, less production energy can be used with increased production efficiency, and fuels themselves can burn cleaner due to better filtration.

Research on Nanotechnology Life Cycles

Prospects for the use of nanotechnology also extend into the realm of environmental protection and remediation. Karn stated that by 2008, the total global demand of nanoscale materials, tools, and devices is projected to be about $29 billion (Business Communication Company, 2004). To ensure that universities and research centers in the United States can perform the highest quality research in this rapidly growing area of technology, the EPA has plans to fund research in the areas of treatment and remediation, environmental implications, and the health and environmental effects of manufacturing nanomaterials, exploring topics such as toxicology, fate and transport, and bioavailability and bioaccumulation.

The EPA is especially concerned with the study of manufactured nanomaterials and their life cycle aspects, the role of industrial ecology, toxicology, and exposure. One EPA project examines the full life cycle of nanomaterials and researches the following:

1. Where are the impacts of products that have nanomaterials in them?
2. Where in the life cycle are their impacts going to fall?
3. Are there any impacts in the use stage like automobiles; the disposal stage, like electronic equipment; or the extraction stage like some of our mining endeavors?
4. How will the move to nanotechnology change a material's flow within a particular sector?

Scientists need to look where in the full life cycle the impacts from these products and processes may occur. Nanotechnology can offer direct beneficial applications to the environment, and society can take advantage of this for both environmental and sustainability benefits. However, it is important to remember that this evolving technology can also have indirect effect on the environment. If solar paint has a persistence problem in the environment, but avoids the pollution created by fossil fuels, or if super-strong materials have similar problems, but help us to avoid the pollution that comes from whole communities being destroyed by a chemical spill, then researchers must be careful to weigh the implications concurrently with the benefits of developing nanotechnology applications, said Karn.

If solar paint has a persistence problem in the environment, but avoids the pollution created by fossil fuels, then researchers must be careful to weigh the implications concurrently with the benefits of developing nanotechnology applications.

Nanotechnology has direct beneficial applications for medicine, public health, and the environment but it also has peripheral effects that can impact the environment, both within the human body and within the natural ecosystem. While taking advantage of this new technology for health, environmental, and sustainability benefits, the U.S. government also must examine the risks concurrently with development of new applications.

Nanotechnology, Human Health, and Medicine

NIH is supporting the development of effective, high output, informative, and less costly systems and has identified several areas for research in nanomedicine, which it defines as integration of nanotechnology and medicine. NIH is planning to support the research on biological molecules such as proteins, DNA, and RNA as well as research into how these molecules interact with each other and with environmental agents. One of its primary objectives is acquisition of a comprehensive database and development of quantitative nanomaterial measurements, said Olden. NIH researchers and independent scientists have identified the need for mathematical and analytical tools to quantify manipulations and interpret measurements of nanomolecules. Thus, mathematicians and chemists need to get involved in the effort to understand what these variations, changes, and readouts mean to biological processes.

Once an early biomarker of a disease or dysfunction is identified, then scientists can use targeted pharmaceutical or gene therapy to correct the faulty components. Additionally, NIH is planning to develop very sensitive detection systems that would be able to detect a single cell or a few cells that are diseased or perturbation of a pathway. Once an early biomarker of a disease or dysfunction is identified, then scientists can use targeted pharmaceutical or gene therapy to correct the faulty components, noted Olden.

Pebble Chemistry as an Example of a Medical Application of Nanotechnology

Nanotechnology, especially as it is applied to biological systems, is an intricate, interdisciplinary field, said Martin Philbert of the University of Michigan. Researchers have begun to build nanoparticles that are intrinsically biocompatible, that do not alert the immune system to their presence, and that can contain a variety of highly functional and highly specific elements that might be toxic but which are shielded from the biology by a shell. One example, the Probes Encapsulated by Biologically Localized Embedding (PEBBLE) was created as a nanoparticle platform with multifunctional capabilities.

Generally, PEBBLEs are fluorescent dyes that are encased in a molecular shell, protecting cells from the dye and protecting the dye from cellular degradation or manipulation, stated Philbert. The presence and intensity of the fluorescence is directly proportional to the pathway or molecule that the sensor is measuring.

Currently, they are being used in research facilities and offer researchers the ability to measure real-time binding, translocation, and molecule production, instead of the widely used tools available for steady-state or endpoint measurements. However, they may have use for a wide range of research and medical needs.

Pebble Chemistry

The simplest iteration of a PEBBLE is a polymer shell that contains a fluorescent molecule, ranging in size from 20 nanometers to 600 nanometers in diameter. One 20 nanometer particle occupies I thousandth of a human motor neuron in the anterior horn of the spinal cord; thus, cells have a potentially tremendous capacity for the retention of a wide range of PEBBLEs sensing and reporting on a variety of cellular pathways and processes.

Philbert stated that the analytical chemistry that can be performed within the conserved space of the cell is much more sophisticated. A dye can be incorporated that changes with time as a function of the analyte of interest. The system can also be constructed in a more complex fashion using ionophores. When irradiated with the appropriate wavelength of a laser, light is emitted and can be measured, allowing for an amazingly useful property where the measurement system is completely calibratable and the fluorescence is reversible, said Philbert.

PEBBLEs as Stepping Stones to Understanding Cellular Processes

One aspect of nanoparticle research has focused on the use of nanoparticles as sensors of cellular activity. This can include extra cellular measurements, such as the presence of oxygen or nitrogen radicals in the extra cellular space and the presence of the molecules in the blood and tissue, or measurements at the subcellular level. Protein movement, oxidation or reduction reactions, and the production of super oxide ions within the cell may be tracked by the use of fluorescent nanoparticles. There are also design prototypes for measuring lead and mercury, and soon, according to Philbert, the ability to measure phosphorylation and other chemical reactions inside the cell will be possible.

The light emitted from individual nanoprobes introduced into the volume of the cell can be viewed using confocal microscopy. The concentration of oxygen at each individual point is reported by a dot in the co-focal viewing field when observing the cell. The intensity of the light emitted by the dot indicates the concentration of oxygen at that point in the cell. With this new look at the cell, Philbert suggested that our assumptions about how oxygen enters the cell, diffuses through the cell, and is used in the cell may not be entirely correct.

Nanoparticles can also be used to track the movement of particular cellular compartments. Endosomes can take up the nanoparticle, and depending on surface charge, size, and other characteristics, transport them to the apical surface or the basal lateral surface of the cell.

The reason why scientists have attempted to go smaller and smaller is that they were trying to find out if there is a point at which the sensors that are being designed and placed into systems to measure biological activity begin affecting or changing the very processes they are intended to measure. Once a molecule is below 100 nanometers, it begins interacting with proteins, potentially signaling proteins, and can result in entirely unintended consequences, which is why PEBBLEs have been kept close to 100, and less than 200, nanometers. Thus while there are many advantages to keeping PEBBLES very small, scientists really do not know yet what the interactions with proteins and other elements of the cell are going to be, said Philbert.

With this new look at the cell our assumptions about how oxygen enters the cell, diffuses through the cell, and is used in the cell may not be entirely correct.

PEBBLEs on the Path to Understanding Tumor Biology

One area of research, according to Philbert, focuses on targeting molecules to tumors for treatment and eradication. Coating a PEBBLE nanoparticle with a magnetically responsive metal allows researchers to target nanoparticles to specific locations within the body and to obtain a better image of tumors from magnetic resonance imaging (MRI). These MRI images reveal a significant contrast enhancement due to the presence of the nanoparticles within the tumor tissue, that is, nanoparticles allow for very good contrast from normal tissue, where the tumor becomes clearly visible to the observer.

Fine resolution of small tumors is not the only possibility within the same nanoparticle. Ruthenium can also be included within the nanoparticle. When a laser is turned on. Sol Gel Ru-DNPs can produce a large amount of singlet oxygen only when it reaches the target, in this case a brain tumor called 9L gliosarcoma produced in the rat brain. Therefore, researchers can kill this clone of the gliosarcoma, which does not respond to chemotherapy or radiation therapy. Coupled with a fiber optic that is only a millimeter in diameter, the tumor can be stopped or at least reduced in size with time. Because the 9L gliosarcoma is extremely aggressive, if one or two cells are left, after some time the tumor will re-grow. However, with a single injection and only 10 minutes of laser irradiation.

The ultimate goal of this research is to create particles that are not going to interfere with the normal biology of the organism and are going to have very high therapeutic index. Nanoparticles used in tumor killing experiments are larger than 100 nanometers; thus, they do not cross the blood-brain barrier (BBB), and therefore are excluded from healthy brain tissue, said Philbert. These nanoparticles penetrate into the tumor only when a tumor changes the porosity of the blood vessel. This process of BBB penetration takes approximately 5-10 minutes and once clearance through renal and other mechanisms is considered to be complete, the laser is turned on. The nanoparticles are delivered systemically and then pass through the BBB before a laser is turned on to activate the killing activity, but further research will be need to understand what happens to these particles as they circulate through other parts of the body?

Approximately twenty different kinds of these nanoparticles have been created and researched in vivo in the Philbert lab alone, all of which have shown effects on the reticular endothelial system and the kidneys. These nanoparticles are discarded as the ultimate goal of this research is to create particles that are not going to interfere with the normal biology of the organism, yet have a very high therapeutic index. Philbert suggested that many of the products that can be created will ultimately not be deployable because early safety testing will show them to be completely unfeasible from a toxicological standpoint.

INDEX

INDEX